LUCIEN GREGOIRE'S

WHITE LIGHT   DARK NIGHT

THE REVOLUTIONARY LIFE OF JOHN PAUL

to

Giovanni Paulo Luciani

beloved father of

Giovanni Paulo I

AuthorHouse™
1663 Liberty Drive, Suite 200
Bloomington, IN 47403
www.authorhouse.com
Phone: 1-800-839-8640

© 2007 George Lucien Gregoire. All Rights Reserved

No part of this book may be reproduced, stored in a retrieval system or transmitted by any means without the written permission of the author.

First published by AuthorHouse 9/12/07

ISBN: 978-1-4343-0692-0 (sc)
ISBN: 978-1-4343-0693-7 (dj)

Library of Congress Control Number: 2007903343

Printed on acid free paper in Bloomington Indiana, United States of America

A compassionate priest, *"Yes, he was all you say he was. My hope for a more just Church and a better world died with him."*

An enraged nun, *"You're wrong. He was much more than you claim he was. Yes, for a short time, Lucifer sat on the papal throne. Fortunately, Gabriel and his angels rose up and tore him down."*

Howard Greene, playwright, London Times, *"Like the Davinci Code, 'White Light Dark Night' will infuriate the devout and other believers in the supernatural. But, unlike Brown, Gregoire has the proof."*

Howard Jason Smith, critic, Boston. *A Princely Dream?*
"This book is the compilation of originally published actions of the 33-day Pope and not as they have been republished by Rome. So outlandish and revolutionary - particularly for his time - are many of them, and so ingeniously and beautifully does the author weave them into the record, one could get the impression, from time to time, one is immersed in a Princely Dream. Yet, unlike the Davinci Code, the critics of this book must contend with the proof of the pudding - footnotes - ironclad sources that define the integrity of this work. Pinch yourself. It's not a dream."

Toby Johnson, White Crane, *"In revealing the dark secret that must have haunted him all his life, he forces the transformation of Christianity."*

Foreword

Giovanni Paulo I

The lion is the most courageous of animals. When the hyena pack comes, though it means certain death, the lioness will stand and defend her cubs and fight to the end. Lions are also among the smartest and most vigilant of animals. With each step she cautiously tests the grass before her.

The sheep is the most cowardly of animals. When the hyena pack comes she will run for her life and leave her lambs behind. Sheep are also the dumbest and most gullible of animals. She has no mind of her own and will follow the sheep in front of her over the cliff.

The three stars stand for the three attributes his father built into him: *Compassion, Courage* and *Change*. The six mounds in the foreground represent the six Dolomite Mountain peaks for which he held the speed record when he became a bishop in 1958. He would set his last speed record climbing one of the Italian Alps most difficult peaks just three months before his death.

Thanks

Those who helped me access records of the Vatican Apostolic Library, British National Library, Cairo Museum, Italian judicial and parliamentary files and Italian dioceses and libraries.

Translations – *Morena Ciringione, Maria Cristina Tolomeo*

Cover including illustration: *Roberto Macedo Alves*

Contributors: *Anna Maria Benelli, Joseph Bernardin, John Calvi, Pascal Cantona, Ray Dubuque, Anita Fournier, Harry Faddis, Hans Lubbers, Giovanni Luciani, Guiseppe Luciani, Archbishop Paul Marcinkus, Anna Montini, Nina Moro, Randall Reade, Teresa Roncalli, Bouwien Rutten, Mario Signoracci, Archbishop Bruce Simpson, Mark Tyminski, Linda Suenens, Hans Willebrands.* Anonymous: *the angel Victoria* and *the angel Elisa*.

Some priests and bishops of the Church who made substantial contributions have asked their names not be printed in this book.

Sources

My direct witness is outlined in the preface.

Most press references were widely published and can be found in major newspapers. For events reported only in local newspapers, the specific publication is footnoted in the text and the corresponding reference can be found at the end of each chapter.

Biblical quotations, unless otherwise stated, are taken from the oldest surviving texts. New Testament quotations are taken from the 4th century *Sinaiticus Codex* held by the British National Library in London. Old Testament quotations are taken from the 10th century *Aleppo Codex* held by the Israel Museum or from the 11th century *Leningrad Codex* held by the Russian National Library in St. Petersburg. Modern Bibles have been materially changed from older texts; many of these changes are discussed in this book.

Table of Contents

Chapter 1   The Worst of Children to the Best of Men 1
Chapter 2   The Minor Seminary at Feltre 19
Chapter 3   The Slaughterhouse of the Human Mind 39
Chapter 4   The Worst of People 50
Chapter 5   The Major Seminary at Belluno 56
Chapter 6   His Philosophies and Goal in Life 74
Chapter 7   *White Light* – his ministry 81
Chapter 8   Two Very Different Kinds of Popes 130
Chapter 9   How a Pope is Elected 142
Chapter 10 Strange Events of August 1978 156
Chapter 11 His Papacy 158
Chapter 12 The Strange Visit of Metropolitan Nikodim 168
Chapter 13 *Dark Night* - his death 174
Chapter 14 The Deception 184
Chapter 15 Pauper who would be Pope 189
Chapter 16 Milan 191
Chapter 17 Vittorio Veneto 197
Chapter 18 The Luciani Expedition into Egypt 214
Chapter 19 The God Zagreus-Dionysus 235
Chapter 20 Albino Luciani and General Patton 242
Chapter 21 *Mud in the Street* 251
Chapter 22 The Winning Card 257
Chapter 23 What They Died For 267
Chapter 24 Thoughts 269
Epilogue   Imagine 271
Addendum History of the Holy Trinity 292

Preface

For those of us who remember him, I bring nothing new. But for those of us who have allowed the Church's misrepresentations of what he was all about, who have allowed its falsehoods to distort his legacy, I bring a treasure trove of yesterday.

Under St. Peter's is a slab of granite, "IOANNES PAVLVS P. P. I"

Like its counterpart in Arlington, it too marks an unknown tomb, *The Tomb of the Unknown Pope*. Unlike the tombs of other popes which tell of their lives, there is no inscription other than his name. Not even the period of reign marks his place in time. The Church would rather his life remain a secret, *The Secret Life of John Paul*. The only pope in modern history for which the Vatican has not commissioned a biography be written. *White Light Dark Night* is the only existing biography of the 33-day Pope.

The dialogue herein is both *italicized* and unitalicized. *Italicized* quotes are supported by direct references. Unitalicized dialogue, although representative of what was actually said by real people, is not supported by public record or documentation.

In order to describe events as close to how they actually happened, this book is written in several genres. This should make it more enjoyable to read and keep it from being the dry narrative that most biographies tend to be. Footnotes are at chapter ends.

The missing will

In the fall of 1978, I had scheduled a vacation to visit my friend Jack Champney in the Vatican. The nature of my visit would change when John Paul died. I would console Jack who had been so close to the Pope. A couple days later, it would change again when Jack was killed by a hit-run driver and shipped back home.

Though I had my suspicions at the time, I have no evidence to this day other than coincidence that Jack's murder was linked to that of his long term friend and confidant Albino Luciani.

Nevertheless, I kept my date. When the Opus Dei candidate – the Polish cardinal - rose to power, I grew concerned. I flew to the Veneto country where Luciani had spent his ministry.

I visited Jack's good friend Antonio Cunial who at the time was serving as bishop of Vittorio Veneto. I hoped to secure some of Luciani's records that I might someday put them to writing. He told me that a group from the Vatican Foreign Minister's office had shown up the week before and had taken everything with them. I asked him why he had surrendered the records.

As part of standard indoctrination, an incoming Pope is required to file his will with the Vatican. Although the Vatican Clerk insisted he had secured the will from Venice a few days after Luciani's election, it had been lost in the Vatican offices. After the Pope's death, his lawyer in Venice was asked to send another copy to Rome. At first he complied. A few days later, he sent a message he had discovered Albino Luciani's will missing from his files.

Cunial told me, when confronted, he had called Luciani's lawyer and was told Luciani's will had provided that his records pertaining to his ministries as a priest and a bishop had been willed to the dioceses of Belluno and Vittorio Veneto. Yet, in that the will could not be found, he had no authority to resist and surrendered the records. To this day, Luciani's will has never been found.

The bishop told me something else. There had been a break-in at the local newspaper and some of its archives had been stolen. This did not mean much to me as when he was bishop of Vittorio Veneto most of the more important things he said and did reached notoriety and were recorded in many newspapers. This is why what I have to say about his time as a bishop and as a cardinal is so well supported in this book; direct references from scores of newspapers and other public records that survived the Vatican's attempt to annihilate the controversial life of Albino Luciani.

As a common priest, however, he was not widely known and one would be dependent almost entirely on Belluno's *Corriere delle Alpi* for those twenty years he spoke out against the Vatican on humane issues. I arrived in Belluno too late; another break-in. Also, Bishop Ducoli had given up Luciani's records on request. This part of his life had vanished as if it had never existed.

So for his time as a child, as an outspoken seminarian and as a revolutionary priest, except for some of his more outlandish attacks on the Vatican which reached notoriety, I rely primarily on my direct witness; my personal encounters with the man himself.

I recall each moment of them as if it were yesterday. I relished those times as I witnessed this good man Luciani smiling, grinning, laughing, teasing, joking and then smiling some more.

He went into great detail and spoke hours on end of his days as a teenage troublemaker in the seminary at Feltre and I recount much of that here. He had less to say of his days in the major seminary at Belluno and as a young priest. Yet, he gave me enough to bridge the gap between his time at Feltre and the time he became a bishop where the record reaches firmness once more.

A great deal of what I talk about here is the record of my friend Jack's correspondence and the many conversations he had with me. In a few cases, I have clarified what Jack had to say in his letters. Only one arrived during John Paul's papacy.

Lucien,

Just a short note. Sorry I haven't written for awhile. Since this thing happened I have been running around doing my best to soften the blow for the losers. You know, Baggio, Oddi, Samore, Casaroli and all the others.

Yesterday, I had to round up a couple dozen of them for him. He talked of the problems of the world.
He told them the Vatican's ban on the pill is resulting in poverty and starvation in third world countries and abortions in first world countries. He told them it is morally wrong for us to stand in the way of long term loving relationships between any of Christ's children, whether a matter of race, creed, remarriage or 'whatever'. Also, it is morally wrong for us to stand in the way of a woman's right to minister the will of Christ. He told them a number of other things. He told them in the coming weeks Mother Church will cease to be the cause of the world's problems and rather will begin to be the answer to them.

Sometimes I wish I had been born better at math like you, than theology. I certainly would have never made it as a typist.

See you next month.

Jack

The old box

There is an old shoebox on the top shelf of my closet. I guard it with my life. For it is my treasure, you see. I want to share a bit of it with you.

Dear Mr. Gregoire,

Most people are brought up to believe when Jesus said, *"Where two of three are gathered in my name, there am I in the midst of them."* He meant going to church and listening to what a preacher thinks is right or wrong. I have found that most preachers think of sex as being morally wrong whereas as a matter of fact, as John Paul often said, *"We think of sex as the greatest of sins, whereas in itself it is nothing more than human nature and not a sin at all."* On the other hand, these same preachers think there is nothing morally wrong in using their power over the weakness of the minds of men to deprive millions of innocent men, women and children of basic human rights and dignity under the laws of nations.

They are also brought up to believe that Christ meant we were to build great edifices of marble and gold and pay men to dress up in gowns of silk and satin to offer pomp and ceremony and wafers and wine in His honor. As you say, this is the reason why one of the greatest concentrations of gold in the world is in the Vatican – the reason why it owns the world's largest collection of ancient sculpture and its works by pinnacle artists would fetch billions at auction – the reason why the wine cellars at the Castel Gandolfo and the Vatican contain the most valuable collection of vintage wines anywhere in the world; while one hundred million children are starving to death in the Catholic world alone as I write this letter.

John Paul thought differently. Even as a boy, he thought Christ meant something else when he said, *"Where two of three are gathered in my name, there am I in the midst of them."* The little boy Albino Luciani thought Christ meant, "Helping others less fortunate than ourselves."

My name is Mark. I am eighteen years old and I am about to start university this September reading Religious Studies. I hope to perhaps specialize in Eastern Theology and go on to writing, with maybe a little music and theatre thrown in as I have done a lot of that and really enjoy it. I would eventually like to work for charities, perhaps in the Third World. I want above all things to help people with whatever talents and abilities I may have, and in whatever way I can. No matter what I do I want to help people somehow. John Paul has inspired me to do this.

Thank you so much for writing such a wonderful book. I will treasure it and try to live up to it. Hopefully, we will one day achieve his dream.

Mark

Chapter 1

The Worst of Children to the Best of Men

Albino Luciani was born into dire poverty in a small village in the Italian Alps to a scullery maid and a migrant worker. His mother was a devout Catholic who prayed before crucifixes made of bits of wood. She told him the only path to heaven was on his knees in prayer. His father was a social revolutionary activist who often burned his mother's crucifixes in the stove. He told him the only path to heaven was on his feet helping others.

After he became Pope, his brother was interviewed, *"It was terrible; we even went without shoes most of the year as to save them for the wintertime."*

One day, when he was just six years old, his grandfather told him the bad news, "Albino, today, you believe in both Jesus and Santa Claus. Well," he apologized, "there is no Santa Claus. We've been kidding you."

He remembers crying himself to sleep that night. How could they take Santa away from him? In his dreams he waved goodbye to Santa, but he still had his Jesus and he pleaded, "Please don't take my Jesus from me."

The next day he trudged along the railroad tracks in the knee-deep snow. His shoes were tattered and torn and worn and they did not match and his feet were frozen and they tormented him with each step. Yet he continued on, pausing here and there, filling his pail with scraps that had fallen from the rumbling coal cars. He thought of the day before when he had gazed through the glass store window at the golden crucifix. At three hundred lire a pail he needed three more pails full and he would have enough to buy this splendid treasure for the very best mama in the world. A smirk of a leer tinged his lips as he imagined his papa's frustration when he would try to burn this one and it broadened into a smile as he imagined the look of surprise and wonderment his mama would have when she would open the gift on Christmas morn. He pictured her puzzled expression when she would read the card,

You gave him life; I gave him hope.
Together for a time; we gave him paradise.
Now my quest must end; but for you the work goes on.
The struggle must endure; for the challenge remains.
The hope is still there; his dream must never die.

Santa

The worst of children

The icicles poured like waterfalls from the roof tops all the way down to the walkways beneath them. In the summertime, each house had its own identity - its own personality - red - green - blue - orange. Each had been a tiny splinter in a giant rainbow. But, now, in the wintertime, each was just one of an endless row of crystal figures in an enormous glass menagerie.

The parade of weather-beaten wooden carts moved through the streets of Canale d'Agordo in the Italian Alps as they had every other morning. The snow was heaped so high on each side of the road that for the most part they passed unseen.

Still, the shouts of the barkers broke the stillness of the morning air: "milk - milk - milk," "cheese - cheese - cheese," "lamb - lamb - lamb," "bread - bread - bread," "eggs - eggs - eggs." Their voices echoing through the white capped rocky mountain gorge which engulfed the desolate town. Yet, one had made its way before them - no wares - no barker - no echo. A silent one - a ghostly one.

The cart rumbled along the darkened snow-covered cobblestones in the wee hours of the morning; its chauffeurs, pausing here and there, picking up their ghoulish haul - those of Italy's two million orphans who hadn't survived the wintry night. Only the creaking of the wheels and an occasional thud of a frozen tot broke the quiet of the dawn.

They were orphans because they were the worst of children
- BASTARDS. They called them BASTARDS because they were children who had been born out of wedlock. Nobody wanted them. That is, nobody in their right mind wanted them. Everyone hated them. That is, everyone who went to church. And in those days

everyone went to church. Every priest, every nun, every monk, every devout parent, every brain-washed child, despised them.

Each time their tiny frozen bodies would pass by in the cart, they all thought it to be right. The only hint of compassion now and then, "They are better off dead." Everyone thought there was something holy about it. After all, it was written in their Holy Bible, these were the worst of children – BASTARDS.[1]

That is, everyone except Piccolo, the little boy Albino Luciani. He thought it was wrong. He didn't care whether or not it was written in a book. In fact, he knew it was wrong. And he knew it was wrong because his revolutionary socialist atheist anticlerical father had told him it was wrong.

A make-believe world

I remember it as vividly as if it were yesterday. It was the very first thing he told me as we sat together for the first time on the terrace of the Bishop's castle in Vittorio Veneto in the spring of 1968. "We live in a world of make-believe. Of all God's creatures, only we are creatures of belief. It is that we believe which makes us think we are better than others.

"Take the story of the young American who went to the cemetery and placed flowers on his mother's grave. While he stood in silence, a young Asian came along and placed a bowl of rice on the next grave. 'Dumb Chinaman,' he chuckled to himself.

"Allowing time for his visitor to pay his respects, he asked him, 'When do you think your father is going to come up and eat the rice?' The young man, thought a moment and replied, 'The same time your mother comes up to smell the roses.'

"It is also this tendency to believe that causes us to become victims of each other. It is what enables a few to control the minds of many. It is why, of all animal species, only humans develop hatred toward their own kind. It is why, of all animal species, only humans annihilate entire populations of their own kind.

"It is that we are creatures of belief that makes us easy prey not only of leaders of nations, but of churches as well. It is why we take the word of the preacher over the word of Christ Himself.

"When Christ said, *'Thou art Peter and upon this rock I will build my Church,'* He did not mean buildings of stone and granite or a wealthy organization made up of places of worship with altars of marble and gold and jewels surrounded by elegantly robed men chanting in vain repetitions. Christ meant His followers were to practice His principles and that is all He meant by His Church. Despite what preachers might fool their congregations into thinking, Christ tells us that going to church has nothing to do with salvation.

"Like all men," the good bishop Luciani told me, "there are three kinds of preachers. There is the scarcest of them all, a few who truly try to carry out Christ's message. Then there are those that want not much more than to fill their pockets with gold. Then there is the third kind, the most dangerous kind, and unfortunately the most prolific kind. Those who want to control the minds of men. Even though Christ could not have been more explicit in what He had to say, there are preachers, so scrupulous in their drive for power, that they lie to their unsuspecting followers about what Christ really had to say about the word Church."

The little known bishop of the remote mountain province continued, "Christ's sole requirement for salvation is twofold, *'Love thy neighbor as thyself,'* and *'Sell all that thou hast and give to the poor.'* Not once in His ministry does He ask us to fall down on our knees and adore Him. There is no message more explicit than is Christ's account of His two archenemies - BIGOTRY and GREED; which today, in direct defiance of Christ's instructions, are the fundamental driving forces of our Christian society.

"So that everyone would understand Him - no matter how dim - on this very important point of His testimony, Christ is explicit as He makes clear the Church He intended was not to be one of stone or a place of worship. Asked if Herod's Temple of Jerusalem was His Church, He answered, *'The Kingdom of Heaven is not in buildings of wood and stone, it is in your heart, it is in our compassion for others.'* Christ tells us that our relationship with Him is a private and direct one - it requires no preacher or middleman - it should not be the public display that today's Christians and the Pharisees of His time make of it."

The bishop Luciani backed up what he had to say with Christ's sacred testimony in *Matthew 6,* "*'And when thou prayest, thou shalt not be as hypocrites are; for they love to pray standing in the synagogues and in public places, that they may be seen and heard of men. Verily, I say unto you, They have their reward in this life. But thou, when thou prayest, enter into thy closet, and when thou hast shut thy door, talk to me in secret; and ye reward will be in the next life. When ye pray, speak to me, do not ask of me. And use not vain repetitions as the heathen do; for they think they shall be heard for their speaking in public places . . .'* ²

"Christ knew his followers would become the very people he so often condemned in His ministry. We, today, are the HYPOCRITES - the Pharisees, He so often condemns in the gospels." Then in a tone of great anger, the bishop added, "And the greatest of all of HYPOCRITES are those preachers who claim to 'love' a certain group of people, yet, spend all their energy in the public forum trying to deny that very same group of their rights.

"Believe me, Lucien, denying any of God's children, no matter how scorned by doctrine, basic human rights and dignity under the law is not loving them. It is hating them; HYPOCRISY of the very first rank." ³

Luciani told me of the first time he had missed Sunday mass. "I had just turned eleven and was as poor as a church mouse and often went hungry myself. Yet, I did have a mama and papa to take care of me and love me. They were away visiting a sick friend on that sub-zero Sunday morning when I made my way to church with my fellow Christians. We passed a dozen orphans begging in the streets. Most of them were orphans because they had been born out of wedlock; the reason why they were barred from church as it had been God the Father's sacred testimony in *Deuteronomy 23, 'A bastard child shall not enter into the congregation of the lord.'* It was this 'holy' testimony of his 'God' that first made me realize what a monster and a liar Moses was.

"It was this 'holy' testimony of his 'God' that first made me realize the Old Testament was not the word of God. As a matter-of-fact, it was not even inspired by God; it was obviously inspired by the hatred and greed of men.

"It may have been the intense cold that inspired me, but nevertheless, I turned around and hurried back to my house and quickly cooked up a caldron of soup with all the vegetables and lentils I could find and although it meant that we would go without them ourselves for several days, I took it to the orphans and placed it in the snow in the midst of them. For the first time in my life I realized what Christ had meant when He said, *'Where two or three are gathered together in my name, there am I in the midst of them.'*

"By that time church was over, so I had missed mass - a mortal sin in those days. I decided to go into the church to ask forgiveness. But I forgot they locked the doors outside of service hours to keep the orphans from coming in to get warm.

"It was then - at that very moment - when those doors would not open - I realized what Christ had meant by the word 'Church'. It was then - at that very moment - when those doors would not open - I decided to become a priest. It was then – at that very moment - when those doors would not open - I decided I would change the Church back to what Christ had intended. It was then - at that very moment - when those doors would not open - I first realized my devout mama was a sheep, and my socialist revolutionary papa was a lion. It was then - at that very moment - when those doors would not open - I began to shed my wool, and groom my mane.

"I hoped the scolding I would receive when my parents returned would not be too harsh. I underestimated the wrath of my devout mama who knowing that I had violated God's will, took me into the bedroom and left me there on my knees pleading for God's forgiveness for having broken His sacred law.

"Later, when my papa returned, he pulled me up off my knees and hugged me. He told me that what I had done was wonderful. He told me to ask Christ to forgive my mama for having scolded me. I should not think ill of her, as she was caught up in religion - Christianity - something he often referred to as the *Opium of the Masses*. Drugged with belief, she is unable to judge what is truly right or what is truly wrong.

"So, although it was my mother who had taught me the idolatry of Christ, it was my father who taught me the reality of Christ; it was my father who had taught me right from wrong. Although I loved her dearly, my mother was too caught up in the *Opium of the Masses* to know right from wrong."

The best of men

Many years later on a chilly autumn evening, I sat by the fireplace in an overstuffed armchair. I reached for the newspaper and read: *London Times,* September 21, 1978, Vatican City. As he has on other occasions, after a general audience, yesterday, John Paul called for assistance from his listeners. A young boy came forward.

The greatest of sins

The Pope asked the boy, *"What is your name and how old are you?"*

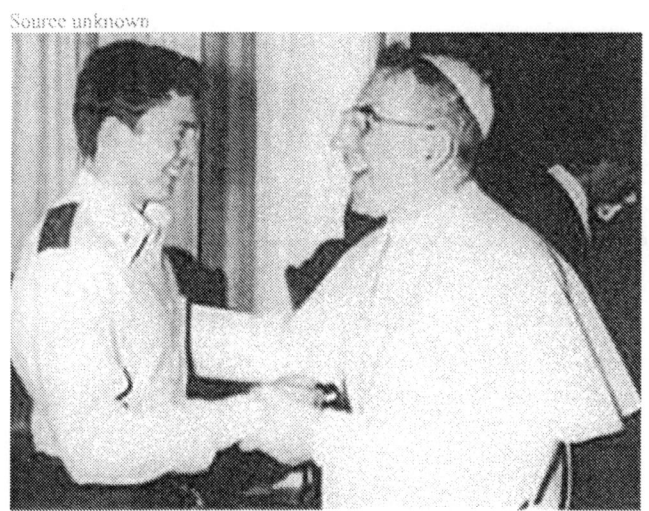

John Paul and Anthony

"Anthony." Then with a touch of pride, *"I am sixteen."*

John Paul asked, *"Good. Now, tell me Anthony, what is the greatest of sins?"*

The boy hesitated, looked sheepishly around the room, hesitated again, and finally stammered, *"Why, I suppose, sex?"*

Smiling, the Pope apologized, *"Sorry, to have put you on the spot. Yes, sex. We think of it as the greatest of sins, whereas in itself it is nothing more than human nature and not a sin at all. I will give you another chance. Now, Anthony, think, what is really the greatest sin of all?"*

The boy thought for a moment or so and answered, *"I guess, murder?"*

"Well, you are getting warmer. But, what I am looking for is the cause. Not the result."

The boy remained silent. The Pope told him, *"Anthony, the greatest sin of all is hatred - hatred of other kinds of people and hatred of those who live their lives differently. And hatred usually goes hand in hand with its partner, greed."*

The boy interrupted, *"Now, that I think of it, you're right. Hatred and greed have been at the root of all of the grief of mankind, the countless wars of ethnic cleansing, murdering and destruction of individuals and, at times, entire populations."*

"Why do you suppose you didn't think of it when I first asked you?" the Pope asked.

"I don't know; maybe because it is not a commandment." The boy replied.

"Why do you suppose Moses left hatred, the greatest of sins, out of his commandments?" John Paul continued to probe the boy.

"Well, I guess, he intended to leave it out or he made a mistake." Anthony responded.

"You are right. But, nevertheless, unlike Moses who was a man and could make mistakes and be driven by personal motive, Christ, who is God, makes no mistakes and is impartial. In Matthew 19, Christ is asked, 'Which of the commandments must I keep, that I shall have eternal life?'

"In His reply, Christ makes no mention of Moses' first four commandments which require adoration of the God of the Old Testament: 1) I am the one God, 2) no graven images, 3) don't take the name of the Lord in vain, and 4) keep the Sabbath. Unlike Moses' God who tells us the only path to Heaven is to fall down on our knees and adore Him, Christ tells us falling down on our knees and adoring Him isn't going to get us anywhere.

"In Christ's response to the question as to which of the commandments are necessary for eternal life, Christ retains five of Moses' requirements: 5) honor thy father and mother, 6) thou shalt not kill, 7) thou shalt not steal, 8) thou shalt not commit adultery, 9) and thou shalt not bear false witness.

"Christ specifically excludes Moses' tenth commandment, 10) 'Thou shalt not covet (desire to take from) thy neighbor his property, including his house, his wife, his slaves, his ox, his ass.'[4]

"Christ knew that Moses' intent was to protect the right of one man to enslave another and to hold women, as mere animals, as man's property. Christ knew that was wrong.

"Finally, Christ adds two of his own - the two that Moses, either through error or intent, left out. 'Thou shalt not hate' - the reciprocal of 'Love thy neighbor as thyself' and 'Thou shalt not greed' - the reciprocal of 'Sell all that thou hast and give to the poor.' Christ's message is clear. The only way we can get to heaven is to help our neighbor no matter who he is or where he is. A lifetime on one's knees in church praying for special favors and selfish miracles isn't going to get anyone anywhere. A single act of kindness to a beggar in the street or a child in Uganda could reap one the Kingdom of Heaven."

The Pope's voice took on a tone as if he were revealing a great secret, "Anthony, this is what Christ meant by the word 'Church'. Take this with you wherever you go. Tell it to others. Spread the word. More than any cardinal or bishop in robes of silk and satin who holds golden chalices up for all to see, you will truly be the successor of Peter. Much more than I will ever be, you will truly be the Vicar of Christ on earth."

A letter from Jack

A few days later I opened a letter from my friend Jack. Jack had served as John Paul's secretary when he had been Bishop of Vittorio Veneto - the reason why I had the good fortune to have met the great man on my early visits to Europe. *"Dear Lucien: Yesterday, John Paul assembled the two dozen or so Vatican cardinals who hated what he stands for so much they had refused to vote for him on the recount that had been taken to render his election unanimous.*

"He talked of the problems of the world. He told them 'The Church's ban on contraception is the driving force behind the spread of disease, poverty and starvation in third world countries and abortions in first world countries.' He told them 'It is morally wrong for holy men to stand in the way of long term loving relationships between any of God's children, whether it be a matter of race, creed, remarriage or homosexuality.' He told them

'It is morally wrong for holy men to stand in the way of a woman's right to minister the will of Christ.'

"He told them many other things. Then he told them one thing more. He told them that in the coming weeks, *'Mother Church will cease to be the cause of many of the world's problems and rather will begin to be the answer to them . . .'* Jack."

The rich and the poor

A week or so later I picked up a newspaper and read: *Associated Press,* September 28, 1978, Vatican City. In a general audience yesterday, the Pope talked of the Church's responsibility to help control the world's population. *". . . I have been discussing birth control for about forty-five minutes. If the information I have been given, the various statistics, if that information is accurate, then during the period of time I have been talking, over one thousand children under the age of five have died of malnutrition. During the next four hours while you and I look forward with anticipation to our next meal another five thousand children under the age of five will die of malnutrition. By this time tomorrow thirty thousand children under the age of five, who at this moment are alive, will be dead of malnutrition. God does not always provide. It is our sacred responsibility to provide,"* the Pope raised his voice, *"and we will provide now."*[5]

The newly elected Pope continued to talk of poverty and starvation in the world, his tones, strong and precise, each word in perfect diction; his voice unwavering in its objective. He spoke in simple words so that the youngest of the children present could understand what he had to say. As he had on many occasions as a bishop, he threatened the hypocrisy of the Vatican treasures.

"This morning, I flushed my toilet with a solid gold lever. At this moment, bishops and cardinals are using a bathroom on the second floor of the Papal Palace which trappings, I am told, would draw more than fifty million dollars at auction . . . Believe me, one day, we who live in opulence, while so many are dying because they have nothing, will have to answer to Jesus as to why we have not carried out His instruction, 'Love thy neighbor as thyself.' We, the clergy of the Church together with our congregations, who substitute gold and pomp and ceremony in place of Christ's

instruction, who judge our masquerade of singing His praises to be more precious than human life, will have the most to explain."

Standing all the while, John Paul spoke for over four hours about poverty, pausing only to take a sip of water now and then. Alternating between languages, he moved back and forth between Italian and English and then in and out of a dozen other languages including a half-dozen African dialects with no discernable difference in efficiency among them. At times, he spoke in tones of simple conversation and at other times in towering eloquence.

Search for the truth

Finally breaking from his discourse, he asked, *"Can one of the children come up to help the Pope?"* A young boy stepped forward and John Paul motioned him to the microphone. He asked, *"What is your name. What grade are you in?"*

The boy stammered, *"Daniele. I am in the fifth grade."*

The Pope followed, *"Now, do you want to stay in the fifth grade or would you rather go on to the sixth grade next year?"*

The boy startled the Pope, *"I want to stay in the fifth grade. If I go on to the sixth grade, I will lose my teacher."*

John Paul and Daniele

Smiling to the crowd, the Pope responded, *"Well, this boy is different than was the little boy Luciani, for when I was in the fifth grade, I would say to myself, 'Oh, if only I were in the sixth.' And, when I was in the sixth grade, I would say to myself, 'Oh, if only I were in the seventh.'"* Turning to the boy, *"You see, Daniele, we have within us a need to make progress - to move forward. Only with progress can we find the truth. We started out living in caves and then progressed to huts and now we live in homes with modern kitchens and bathrooms and telephones.*

"But, more importantly, has been our progress in accepting our fellow human beings as we do ourselves. Our enemy is what has gone before us. We wrongly assume that our ancestors were smarter than we are. We wrongly accept that what they wrote in their books is the truth. We assume that what they had to say was right and this misconception pulls us backward, instead of forward. Yet, as a matter-of-fact, we are much smarter than were our ancestors. As each generation comes forward, it benefits from all the knowledge and truths that have been accumulated by all the generations that have gone before it.

"Tell me, what did God do on the first day of creation?"

The boy looked up at the Pope with a puzzled frown.

"Why, he divided the waters that were there to create Heaven."

The Pope followed, *"And, on the second day?"*

Daniele, *"He gathered the waters together to allow the dry land to appear and He grew grass, trees and flowers."*

John Paul nodded a look of agreement, *"Very Good. Now, how about on the third day?"*

Daniele, *"God hung the sun and the moon and the stars in the heavens to give light."*

The Pope smiled, *"You have a good teacher. No wonder you don't want to leave her. Nevertheless, when Moses told us the story of creation, he was unaware the earth was round and rotating on its axis and controlled by its sun. As a result, he told us God had told him the earth was flat and God hung the sun and the stars in the heavens the day after He had created the earth together with its vegetation.*

"He told us this because he didn't know at his time it is the sun that is the center of our solar system and controls all life on our planet. It is because he didn't know the facts - the truth - that

White Light Dark Night

Moses told his followers that God had told him God had created vegetation the day before He hung the sun in the heavens. Today we know the facts - the truth - no vegetation can exist without the process of photosynthesis which is a product of the sun. As a matter-of-fact, the earth, itself, could not exist without its sun."

Daniele, looking at the Pope with a hint of a smile, corrected him, *"But, God would have told the story of creation as people at the time believed the world to be."*

The Pope acknowledged the boy's response. *"Yes, in the seventeenth century when Galileo proved the organization of the universe - when he proved the story of creation as told by Moses was not the truth - Pope Innocent X declared that God would have told the story of creation in the way that people understood the world to be at that time. Everyone believed him. The reason they believed him was that he was the Pope and at that time the Pope controlled the minds of men. Today, many still believe Pope Innocent's proclamation; that a God, who gave us the commandment, 'Don't bare false witness,' lied about the organization of the universe in His very first words to man.*

"But, of course, I too am Pope. In addition, I have at my disposal the immense abundance of knowledge the world has accumulated since Innocent's time. I ask myself, 'Would God have had any reason to have lied?' The obvious answer to me is that God had no reason to have lied. As a matter-of-fact, God had great motive to have told the truth. For had He told Moses the truth, had God revealed the true organization of the universe to Moses, it would have proved that He was the true God.

"It would have explained a great mystery of that time, as to how it was possible for the sun to rise in the east each morning and descend in the west each evening? How did it ever return to the east? So God had great motive to have told the truth. Yet, Moses, in convincing his people that he had talked to God, inspired his people to take the Promised Land which made him the wealthiest man of his time, save Pharaoh. So Moses would have had great motive to have lied.

"Daniele, now listen very carefully to what I am about to say to you. I want you to keep it with you always." The Pope paused and spoke slowly and decisively, *"Daniele, I do not believe that the same God who has endowed us with reason and intellect has*

intended us to forego their use; that we would believe that a God, who had great motive to have told the truth, lied; and that a man, who had great motive to have lied, told the truth.

"In his day, Moses thought he was telling the truth. But, today we know that he was not telling the truth because, at his time, he did not know the truth. Daniele, if we are ever to know the truth, we cannot start with falsehoods. We must start with what we have determined to be the truth to a given point in time and go forward. As each generation comes and goes we grow closer and closer to the truth. Believe me, Daniele, the truth does not lie in the past; it lies in the future; and it is your job to help society in its struggle to find it."

John Paul went on, *"Daniele. Let me tell you a story. In America, for two-hundred years, Negroes were taken from their homeland and placed in bondage and what is worst of all is that the white church-going Christians thought it was right, that it was somehow holy, that it was the will of God. And they thought this way because, at the time, they were in the fifth grade and wanted to stay in the fifth grade because they could not give up their teacher. Their fifth grade teacher, at that time, was the God of the Old Testament. He told them in His Tenth Commandment, 'Thou shalt not covet (desire to take from) thy neighbor his ... slaves,' that slavery was right. That slavery was the way of the Lord.*

"Then, one day, a man named Lincoln and others like him came along and they said to themselves, 'We shall go on to the sixth grade.' There they found a new teacher, His name was Christ, and Christ told them slavery was wrong." 4

The boy whispered, *"I never thought of it in this way."*

John Paul continued, *"Through the centuries, Mother Church has had the same problem the American Christians had back before Lincoln's day. Back in Moses time, the Israelites, in an unprovoked attack, slaughtered hundreds of thousands of their peaceful Canaanite neighbors including all of their children and thought it was right. Back in Christ's time, they would stone unwed mothers to death and thought it was right. During the Crusades, we murdered millions of helpless Muslims and Jews and thought it was right. Then for six centuries, we tortured and murdered countless innocent people in the Inquisitions and thought it was right. In the World War, your country sided with Germany in its quest for the superiority of the Aryan race and the annihilation and subordination of other races. Again, we thought it was right because we were a Christian nation and the Vatican told us it was right. And, until very recently, we would treat born-out-of-wedlock children as outcasts of society and, once more, we thought it was right. Even today, we continue to persecute many kinds of people who live their lives differently and we continue to think it is right because of what someone once wrote in a book.*

"Daniele, not too long ago when I was your age, every town and village in Italy had a cart that went about the streets each morning picking up the frozen bodies of the orphans who had not made it through the winter nights. At the time, all we good Christians thought it was right because Mother Church told us it was right. But now, just a few generations later, we all know it was wrong. It was an atrocity.

"Just two months ago, the first artificially inseminated child was born in England. The latest polls show that more than ninety percent of Catholics condemn the child. The reason they condemn the child is because Mother Church condemns the child. She tells her 'sheep' that today's 'test-tube' babies are what children born out of wedlock used to be. They are the new BASTARDS of society. So even, today, Mother Church does not know right from wrong.

"Daniele, you too must make progress or you will never know right from wrong. You will never be able to help your fellowman in his search for the truth. You will never be able to make your contribution toward making this a better world to live in for those who come after you."

Daniele, *"Now I know why I must go on to the sixth grade."*

"Yes, Daniele, you have your sacred commission. You must go on to the sixth grade, then on to the seventh, and then on to the eighth. You must go forward, always advancing, never looking back. Progress and change must be your guiding ambition. So that someday the whole world will come to know the truth. So that the day will come about when all men and women will treat each other equally as Christ had commanded, 'Love thy neighbor as thyself.' Not as they have in the past, just because someone once wrote something in a book." [6]

Women ordination

On *CBS News* that same evening: Vatican City. Pope John Paul held a private audience with a delegation of bishops from the Philippines led by Jamie Cardinal Sin this morning. One of the bishops challenged the new Pope on doctrine concerning rumors a woman might soon be ordained. John Paul told the bishop, *"When I was a teenager my father made me promise that I would live my life in imitation of Christ, and I have kept that solemn promise. Each time that the fork in the road has come up, often only minutes apart, I have asked myself, 'Now, what would Jesus have done in this case?' And I have often pondered the possibility as to how much better the world would be if everyone were to do this."* The Pope asked the bishop, *"Now, what do you think Jesus would do in this case?"*

When the bishop remained silent, the Pope told him, *"It makes no difference what was written by self-serving men for yesterday, all that counts is what Jesus would do today."* John Paul then reached for a microphone and raising his voice he told the delegation, *"Never forget that God is more our Mother than She is our Father!"* 7

These would be the last of his words the press would record. The very next day, I picked up another newspaper and read: *Associated Press*, September 29, 1978, Vatican City. *Just thirty-three days into his pontificate, Pope John Paul died last evening... Vibrant and on the job to the end, he was sixty-five... the only Pope in history whose death was unwitnessed... On hearing the news, Cardinal Benelli of Florence called for an autopsy... Born of a social revolutionary atheist father who had placed him in a seminary at the age of eleven with the commission to bring change to the Church... What would have been John Paul's papacy is perhaps best defined by the central message of his acceptance speech in the Sistine Chapel, August 27, 1978, "... We must rise up the courage that is within us and set aside the prejudices that have been built into us by our Christian forefathers and together we will muster the strength to lift those restraints that have been unfairly placed upon the everyday lives of so many innocent people by doctrine... for God-given human life is infinitely more precious than is man-made doctrine..."*

White Light Dark Night

Five days later, I read: *Associated Press,* October 4, 1978, Vatican City: *The coffin, a plain pine box as reserved for paupers, was hemmed in by the princes of the Church in their rich and elegant attire . . . Cardinal Leon Joseph Suenens, Archbishop of Brussels, gave the final tribute for his dear friend.*

". . . Like a shimmering white light, he rose up from the mud in the street and left no one untouched.

For those of us at the top, from heads of churches, to leaders of nations, to those of great scientific achievement, he was the Enlightener - the Imitation of the Holy Ghost.

For those of us at the bottom, from the poor, to the homeless, to the handicapped, to the oppressed, he was the Redeemer - the Imitation of Christ.

But, above all, he was the best of men." [8]

So was the beginning and the end of Albino Luciani. Now, witness the whole of him. This is the Testament of John Paul I.

Albino Luciani = White Light

[1] pg 3 BASTARDS. Moses condemnation of born-out-of-wedlock children is the reason the word continues to have such a terrible connotation today despite the fact that the related stigma is now dead. In 1973, Paul VI motioned to make Luciani a cardinal. Luciani sent a message that unless the Pope reversed *Canon Law*'s condemnation of bastards, he would refuse the *red hat*. Paul complied. Yet, even today in conservative Rome, one will not find a priest who is willing to baptize an illegitimate child.

[2] pg 5 *Matthew 6*, This is the translation of Christ's definition of prayer as found in the oldest surviving New Testament texts. Most modern Bibles replace the phrases "*in this life*" and "*in the next life*" respectively with the words "*already*" and "*then.*" The King James Bible goes a step further and takes Christ's words completely out of context by deleting both phrases entirely; both sentences ending in the word 'reward'. All modern Bibles delete Christ's words, "*When ye pray, speak to me, do not ask of me.*" Christ asks that one not ask Him for favors.

NOTE: The author encourages readers to search verses referred to in this book as they appear in various Bibles. The Jewish *Torah*, the Catholic Bible and other versions including Eastern versions and the King James Bible are available in most libraries and on the Internet. Translations of the *Codex Sinaiticus*, the *Aleppo Codex* and other older documents referred to in this book are available from their present day owners.

[3] pg 5 HYPOCRITES. The bishop was most likely referring to causes he had championed in Parliament which had made interracial marriage legal and permitted single persons to adopt children. Yet, he could have been referring to most any kind of oppressed people. Today, evangelistic preachers claim to love homosexuals, yet they spend hundreds of millions of dollars they collect for the poor in advertising aimed at depriving these same people of equal rights.

[4] pg 8 The Tenth Commandment as it appears in the oldest surviving texts and in some eastern Bibles today including the Jewish *Torah*. "*Thou shalt not covet (desire to take from) thy neighbor his property, including his house, his wife, his slaves, his ox, his ass.*" In the twentieth century, the King James Bible and some other versions changed its meaning as handed down by God to Moses on Mount Sinai: "*Thou shalt not covet thy neighbor his . . . servants or employees. . .*"

[5] pg 10 This same text is reprinted in David Yallop's book *In God's Name*. John Paul first said this in a conversation with Cardinal Villot on Sept 19, 1978

[6] pg 15 The boy Daniele interview. This is a partial but relatively substantial text of the release. Abbreviated and edited versions of this interview have been published in other publications including the best selling book in this category *In God's Name* by David Yallop. The Church's releases of this event are brief and edit out most of the 'Daniele' dialogue. It has long been Vatican policy to edit out anything controversial a pope might say before releasing statements to the press. General audiences, however, are attended by the world press.

[7] pg 16 "*God is more our Mother than She is our Father.*" Despite the fact that the word "She" was widely reported in newspapers the next morning, the Vatican, in its official release of John Paul's statement "*God is more our Mother than She is our Father*" changed the word "She" to "He." The Propaganda Machine of the Vatican is the most organized and convincing in the world.

[8] pg 17 *L'Osservatore Romano* 5 Oct 78 'Albino Luciani' translated in English is 'White Light'

Film excerpt audience 27 Sep 78: http://youtube.com search: Albino Luciani

Chapter 2

The Minor Seminary at Feltre

It was a dark rainy dismal afternoon in October of nineteen hundred twenty-three, when eleven year old Albino Luciani climbed into the carriage that would take him to the minor seminary at Feltre; the first stop on a long journey which would eventually lead to Rome.

His father, who had spent a lifetime trying to change the Church from the outside, decided that it could only be changed from the inside. So it was that he committed his son to the task.

In his farewell, the revolutionary outcast of the Veneto country told the boy, "Piccolo, unlike those hypocrites who prance about the Vatican palaces in their magnificent robes of silk and satin with jeweled chalices and rings of diamonds and rubies and gold, you must promise me that you will live your life in imitation of Christ. For Christ would not approve of this masquerade on the part of His earthly representatives. You must play your cards carefully and work hard until that day when at the helm of their ranks you will establish the common dignity of all God's children in the Church. You must find the strength along the way to do what you have to do to bring about a day when all men accept each other as being equal." So it was that his papa together with his little brother, with tears in their eyes, waved goodbye, on that dreary drizzly autumn day, to this *Pauper who would be Pope.*

The road to Feltre had been a painstaking and difficult one for the frail little boy. In the impoverished town of Canale d'Agordo, his family had been the poorest of the poor. There was a reason for this. His father as a revolutionary socialist activist had been a thorn in the side of the Church. As a result, no one in the village would give him a job. He had to travel hundreds of miles away where he was unknown to earn enough to support his family. Also, he spent much of his meager earnings building small sheds for the orphans. As a result, Albino had grown up in a state of borderline subsistence leaving him a fragile little boy. So much so, he acquired the nickname 'Piccolo' - little one - despite the fact, he was of average height.

An anonymous benefactor

One might ask how it was possible his father had raised the substantial sum of money required at that time to put the boy through the minor seminary of Feltre. Private schools at the time, particularly prep seminaries, were expensive - the reason why priests came only from wealthy families. The best a poor boy could hope for was a monk's robe. All that history knows for sure is that the grant was made anonymously.

His father most likely secured the grant from either the Communist Party or the Socialist Party as they were the only organizations to which he had ties that could have possibly come up with that kind of money. The most logical conclusion one can come to is that his father convinced either one or both of these parties, that the Church could only be changed from the inside. He would contribute Albino - they would contribute the money.

This is certainly substantiated by what is known of him today. Through the years, Luciani allied himself with Communists. As a priest and as a bishop he lobbied openly for a redistribution of wealth society - the central core of Communist philosophy. In 1963, he was responsible for the first dialogue between the Communist Party and the Vatican. He arranged a meeting between John XXIII and the son-in-law of the Soviet Premier Nikita Kruschev, Alexis Ajubei.

As an archbishop, he was known to frequent a Communist club in Mestre on the outskirts of Venice. In 1977, an Italian bishop wrote a letter to Enrico Berlinguer proposing an ongoing dialogue between the Church and *Communism*. Luciani was the only cardinal to support the bishop's recommendation. The next year, when he became Pope, he granted his first audience to the Communist Mayor of Rome.

His first private audience with a foreign dignitary was with Metropolitan Nikodim of St. Petersburg who was the central figure in the Christian world in the effort to bring about basic Communist ideology; redistribution of wealth society. Like Luciani, Nikodim was an enemy of the Old Testament, so much so he had coined the phrase the *God Satan* in describing the *God of Moses.*

Just a few months before he became Pope, Aldo Moro announced a union between his Christian Democratic Party and the

Communist Party. Most cardinals condemned the action. Luciani called the union *"a giant step that will eventually bring about the equality of all God's children."*

It was Luciani's close alliance with Communists that led several authors to implicate CIA and British Intelligence in his mysterious death. It was certainly not in the best interest of these capitalistic countries to have the most influential man in the western hemisphere preaching a redistribution of wealth society.

Although one cannot say for certain, it is reasonable to conclude that the Communist Party paid Pope John Paul's way through school.

His solemn pledge

Feltre was a big step up for the little boy Luciani. To begin with, the building had indoor plumbing. In the poor mountain village where he had grown up, none of the houses had indoor plumbing. Going to the bathroom was the worst of times - particularly in the wintertime. That is, it was the worst of times for everyone except the orphans. They would often sneak into outhouses to keep warm - for them it was the best of times.

For this reason most people kept padlocks on them to keep the orphans out. Nevertheless, going to the bathroom would never again be the worst of times for Albino Luciani. Actually, the worst of times for this little boy were not over - they were yet to come.

The bishop told me of the first time he knelt in the chapel at Feltre alone. "It was then at the age of eleven that I began my ministry. The memory of the hopeless struggle of orphans would haunt me for the rest of my days. These children of God would become the central focus of all my energies. It was then, in the solitude of that tiny chapel, as I knelt on my knees, that I made my sacred pledged to Christ,

'I can offer you no great cathedral, no chant, no offering of gold. All I have to give is my promise that I will do what you have told me to do. I know not where this path leads. All I know is that you have told me to take it, and that is all I need know.

"If at its end there is nothing there, it is enough for me that you have allowed me the opportunity to have walked this way. That you

have called upon me to bring about the equality of all of your children.

"No matter what its destiny, I will always cherish having been given this great privilege. I care not if it takes me over the highest mountains, or across the widest seas, or against the armaments of all the armies of the earth, or even through fire. I intend to do this thing with all the strength, vigor and courage that is within me as if the very existence of each and every one of your children depends on me, alone.'"

Thus began the ministry of Pope John Paul I.

The tyrant of Feltre

Visitors to Feltre today can view a display of old notes, the only surviving record that the boy Luciani had ever been there. A dozen or so reprimands; some of them not much more than a scolding or a slap on the wrist, while others forewarned of possible expulsion and even excommunication.

As he had been in grade school, the little boy Luciani was a rascal and a tyrant. The Church's biographical briefs will tell you this. But they will not tell you why. They will lead you to believe he spent his schooldays shooting rubber bands and dipping pigtails in inkwells. Not so. Here, as I have promised, I will tell you why.

When the youngster questioned that Moses talked to a God who did not know the earth was round and joked about Adam as having been a caveman who carried a big club, it brought him detention. It wasn't very long before his classmates came to know him as the *Doubting Thomas of Feltre.*

One day, he surprised the class when he agreed with his teacher. He agreed that the Jews had been responsible for Christ's crucifixion. The teacher made the mistake of letting Albino have the floor. The youngster stood up and asked the teacher, "But why did the Jews want Christ put to death? What motive could they possibly have had to do such a gruesome thing?"

When his teacher remained silent, Albino looked around the class for a possible answer. With nothing but question marks on their faces, he told them, "It is quite obvious why the Jews wanted Christ put to death. They believed fervently in Moses. Even, today, their central scripture, the *Torah,* is limited to the five books of

Moses. Christ spent His entire ministry contradicting almost everything Moses had to say. When the Jews would take the prostitute or the unwed mother outside the city gates and begin to stone her to death, Christ would intervene, *'Let he who is without sin, cast the first stone.'* Only an imbecile would read the five books of Moses and then read the four gospels of Christ and believe that Christ could possibly be the same God, as was the *God of Moses*. Particularly, a God who gave us the commandment, *'Honor thy father and thy mother'"* Even when he would agree with his teacher, it would bring Albino detention.

The strange little boy in the schoolyard

When he had first arrived at Feltre, he had gazed up in wonderment at the bell tower which annexed the school. It was the first time in his life he had seen a building so tall - easily four times the height of the next tallest building in town. Actually, in total mass, it was every bit as big as the school was itself. It made no sense to him that one would build a structure of what had to be monumental cost just to house a bell. After all, a bell twice the size at street level would be heard further away than would a bell of its size from the tower. Why not just make a bigger bell?

It was at that moment, when he gazed up at the bell tower, that he realized how far the Church had strayed from what Christ had intended; that it would use an immense amount of money it collected for the poor, which could be used to build housing for a hundred orphans, to house a single bell. He noticed one other strange thing about the bell tower. There was no lock on its door. Why no lock on the door to the bell tower? He was soon to find out there was another function of the tower.

It was on his twelfth birthday that he found out that there was another kind of BASTARD. He recalls Giovanni in his memoirs, *"He was a frail little boy who spoke with a lisp and waved his hands in a funny little way. All the kids laughed at him in the schoolyard. Then one day he died. No mass was said for him and he was buried outside of town in the village dump. The day after he was buried, the good Father explained to the class that Giovanni had been born bad so the Devil had taken him back. But I knew the priest was wrong. I knew that Giovanni, like all Christ's children, had been born good. I also knew why Giovanni had jumped from the bell tower. He just couldn't take it anymore. That afternoon, as I stood over my friend's grave, I vowed that I would never let anyone laugh at him again."*

As he crept into his teens, Albino soon came to realize that more than half of his classmates were gay despite the fact that few of them displayed effeminate traits as had Giovanni. He knew the reason why. A teenager who contemplated the priesthood had to make the great sacrifice of celibacy.

Under *Canon Law*, all sex outside of marriage is a mortal sin. A straight youth had the option of marriage and could look forward to a life of sex free of sin. So for him, celibacy was a great sacrifice. But, a homosexual teen, could never marry and therefore, if he was devout, and in those days most everyone was devout, he was condemned by *Canon Law* to a life of celibacy anyway. So in choosing the celibate life of a priest, a homosexual teen wasn't giving up anything.

This is confirmed today, as modern population studies have determined an overwhelming percentage of those who consecrate the Eucharist to be either homosexual or transsexual. For example, more than two-thirds of priest pedophilia indictments have been homosexual in nature. Although, it is known that sexual orientation and pedophilia are separate and distinct conditions, the sexual orientation of the predator does determine the sex of the victim. It is quite obvious the priesthood is overwhelming gay. Many priests tend to be quite homophobic because of the conflict between doctrine and their orientation; they come to hate themselves.

In the five years he would be at Feltre, more than a dozen of Albino's classmates would leap from the bell tower. Some of them were homosexual or transgender teens whose identity was discovered by their 'holy' keepers. A couple of them, he recalls, were discovered to have been born out of wedlock and went off the tower on the eve of their scheduled expulsion. A few others developed some form of deformity: one stopped growing and was determined to be a midget, one was crippled in an accident, a few came down with some sort of disease. Each one, with his dream of becoming a priest shattered, went off the tower on the eve of expulsion - the remains splattered across the cobblestones below.

No mass was ever said for any of them. Each one of them, one by one, was put into a burlap sack and buried in the town dump. For it had been Moses' sacred testimony in *Leviticus 21*, *"The lord spoke to Moses saying, whosoever should he be that hath a blemish, whether he be a blind child, or a lame child, or a child with a flat nose (Negro), or a child that is broken-footed, or broken-handed, or a hunch-backed, or a dwarfed child, or a child of disease is not to approach the altar of the Lord."* [1]

So there were many different kinds of BASTARDS in those days, but nevertheless, they were all BASTARDS. They were the worst of boys - BASTARDS.

His rise to the world stage

Luciani's rise to the world stage did not begin when he became a cardinal in 1973. Rather, it began in 1926 when he was just thirteen years old. He had wiggled himself into the job of Assistant Editor of the school newspaper and in his first article he called on the nations of the world to live up to the responsibilities of their copyright laws and place a warning on the Old Testament,
"This is a work of fiction. Keep away from children." Through the intervention of his father, the article was republished in a socialist literary journal and eventually reached all of Europe.

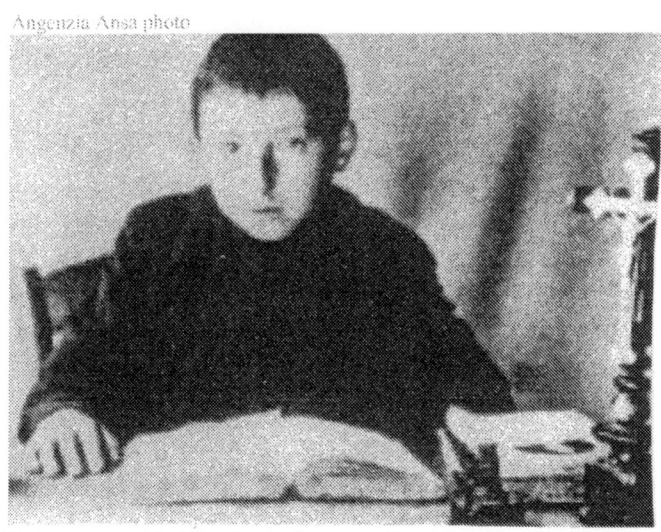

Thirteen year old Albino Luciani peruses the evils of the Old Testament

The article read, "... *The problems of the western world will only come to an end when free nations, in accordance with their copyright laws, require all copies of the Old Testament be prefaced, 'This is a work of fiction. Keep away from children.' The state is derelict in its duty in failing to place, at the very least, a warning on this book, as most people are using it to guide the way they live their lives and this is costing many others their lives.*

"*There is not a single word in the five books of Moses that has been proven to be true. As a matter-of-fact, all of the major claims, without exception, set forth by Moses have been proven to be false. After all, every one of us knows today, the earth is round. Let us not be so foolish as to believe this man Moses who talked to a God who thought that it was flat. It is obvious to any man of good conscience, much of this book was not inspired by God; rather it was inspired by the hatred and greed of evil men.*" [2]

This is still true today. One purpose of copyright laws is to protect the public from what might not be true. Yet, the Old Testament, which many people are foolishly using to guide their lives and enforcing on others, would stand no chance at all before a tribunal. It continues to go unquestioned, despite the fact, as we speak, it is costing thousands of children their lives.

Most of the rank and file laughed, they took it to be a joke. Many of them were Jews. Jews, who would have the opportunity to 'laugh' again, after the last embers of the Holocaust smothered out in 1945.

One must keep in mind that the Jews today use the story of the taking of the Promised Land to convince the world that God gave it to them - thus the Christian world supports the Jews in their ongoing persecution of the Arabs who were its original occupants. Israel sits smack in the middle of the Arab world. Conversely, Hitler employed Moses' Aryan race philosophy in his annihilation of the Jews. There is something in the five books of Moses for everyone; for anyone who has hatred in his or her heart.

Nevertheless, the article drew comments from many noted people, most were critical of the young boy's audacity, some called for his removal from the seminary, others going so far as to call for his excommunication from the Church. On the other hand, one of them, who eventually became a lifelong friend of Luciani, Albert

Einstein, called the child prodigy's article, *"The first bit of common sense to ever come out of the Roman Catholic Church."*

The article was to be the first of hundreds of controversial releases he would make to the press during his lifetime. Many of these support the authenticity and integrity of this book.

The headmaster and rector of Feltre, an aging monsignor, moved immediately for expulsion. The only thing that saved Albino was money. The anonymous benefactor who had made the original grant for his education quadrupled the ante. The money was too much for the diocese to pass up.

The episode made the young seminarian into a celebrity. The picture above appeared in many newspapers, often alongside that of Einstein. The Vatican refused to comment on the issue. After all, it would certainly not be in its best interest to make the matter into a worldwide controversy. If that were to happen, most nations, at the very least, would certainly place some kind of warning on the Old Testament. This would be devastating to the Church, actually to all of Christianity. The very last thing the Church wants is an open discussion in a public forum of the Old Testament versus the known facts. Even at that time, history, medical science, genetics, psychology, archeology, astrology and other sciences would have left the book in shambles. Of course, this is even more so the case today.

Just a few months later, when it seemed the smoke had cleared, Albino struck again. He wrote his first editorial and, again through the influence of his father, although it did not attain the notoriety of the first article, it showed up in an Italian socialist magazine.

"I cannot accept," he wrote, *"that Moses was the holy man the Church and the motion pictures make him out to be. After all, Moses introduced the concept of FASCISM to the western world, that ideology based on a rich and poor society, one in which children are born into without equal opportunity, many of them born into poverty and starvation.*

"In his five books, 'The Fascist Manifesto', Moses talks of God the Father's dream in which the white Aryan male rules at His side and woman is to be held in servitude to man and all others who are different are to be either subordinated, annihilated or cast into slavery. Moses subordinates those he refers to as 'those with flat noses' (Negroes) and 'those who are of physical blemish' (the

handicapped). I am quite dismayed that in all my life I have never seen a black person as black people are not allowed in Italy. We are an entirely white Catholic country because the minds of the voters are controlled by a Vatican that wants to preserve the purity of our Aryan race . . .

"Then there is Moses' horrific taking of the Promised Land in which he puts to the sword all men, women, children and even infants who worshipped gods other than his god. Moses gave birth to what has been more than three thousand years of ethnic cleansing wars which continue to go on even today. The mainstay of his religion is hatred of others who appear to be different.

"Yet, on the other side, we have Christ. He introduces COMMUNISM to the western world, the ideology based on the premise that all God's wealth is to be divided equally among all His children. He dictates a world in which every child has an equal opportunity at a good and healthy life; a far different world than we live in today. Christ introduces a single commandment, 'Love thy neighbor as thyself.'

"In all of His life, Christ commits only one sin, the sin of anger. He so hated the republicans, the money makers, that in a fit of rage He upturns their money tables and throws them out of His Father's house. Christ's other fundamental requirement in the New Testament is that one give up one's material wealth and come follow Him. The mainstay of Christ's religion is love of others no matter how different.

"There are no other philosophies of life. One either believes in the equality of all of God's children or one doesn't. One must choose between COMMUNISM, which is the extreme form of socialism on the left; and FASCISM, which is the extreme form of conservatism on the right. These are the pillars of society, Communism and Fascism. They stand at either end of this rope called humanity.

"The job of the people is to make certain that neither one of these extremes wins the tug-of-war. For that would result in a one-party system and dictatorship. If that were to happen, either the Anarchy of Communism or the Anarchy of Fascism would emerge as an enemy of the people. So it is important we work out something in-between or we will all end up in the drink.

"Yet, one can choose to be closer to Christ or closer to Moses. Although you would never get either one of them to admit it, a good democrat strives toward COMMUNISM and a good republican strives toward FASCISM. That is, the democrat strives toward Christ and the republican strives toward Moses. Despite Christ's overwhelming testimony two thousand years ago, Christianity remains deeply steeped in FASCISM today. It is quite obvious that Mother Church in her support of a fascist state has chosen the word of Moses over the word of Christ Himself." [3]

Albino's editorial was undoubtedly in response to an order by Pius XI, two weeks earlier, requiring all of Italy's children under sixteen be enrolled in the new *Fascist Youth Organization*, which later served as the kindling wood for World War II.

In the spring of the following year, Hitler's best seller *Mein Kampf* was made required reading in the seminary at Feltre as it was in all Catholic secondary schools in Europe. In his book, Hitler defends his plan to annihilate certain populations by quoting passages of the Old Testament, particularly from the books of Moses. At the end of his book, Hitler concludes. *"I believe, today, my conduct in proposing these things is in accordance with the instruction of the Almighty Creator."*

It was that Hitler's philosophy was based so heavily on the ideology of Moses, and that the population believed it to be the word of God, that convinced half of the western world that Hitler was carrying out the instructions of God. This, in the very same way the Israelites, convinced that Moses and Joshua were in fact talking to God, believed they were carrying out God's instructions in their horrific taking of the Promised Land.

In January 2006, Pat Robertson told a television audience, *"The United States should continue to provide arms to Israel to suppress the Palestinians because God gave that land to the Jews."* One would wonder how Robertson knows God gave that land to the Jews; it is quite obvious that if God had wanted the Jews to have the Promised Land then He would have given it to them in the first place.

In retrospect, had the courts of the time acted on the boy Luciani's demand that a warning be placed on the Old Testament, the world may have been spared World War II and the Holocaust.

He tastes of the forbidden fruit

In addition to the required studies of Feltre, which were designed to brainwash its students in *Fascism*, Albino spent much of his time looking at the other side of the coin. Through his father, he obtained a parade of books, most of which had been forbidden by the Vatican.

Among them were Albert Einstein's *Theory of Relativity* which declared the egg came first; the story of *Adam and Eve* was a fairytale. Then there was Darwin's *Origin of the Species* which spoke of evolution, man and ape had evolved from common parents; something that today's DNA advancements have proved as scientific fact. Then there was Gregor Mendel's *Experiments in Plant Hybrids* which backed it up.

There were a great number of books on anthropology, archeology, chemistry, physics, history, and most importantly on psychology - how the mind works. Then there were those on socialism, Marx and Engels' *Communist Manifesto* and Marx' *Das Kapital*. There was Vladmir Lenin's *State and Revolution* and *The Rise of Capitalism in Russia* and a countless number of others.

He found these fathers of the Russian Revolution had tried to bring Christ's teachings into a selfish world.

Communist Ideology	Christ's teachings
All God's children are equal	Love thy neighbor as thyself
Redistribution of wealth society	Sell all thou hast/give to the poor

He discovered Communism was the enemy of the Vatican because it was the great enemy of bigotry and greed.

Finally, there were those that were specifically banned from all seminaries in the Catholic world. There were the Bibles of other religions including Mohammed's *Koran*, the Hindu scripture the *Vedas*, the Buddhist scripture the *Tripitaka*, Martin Luther's *Thesis 95* and an endless number of others. He spent countless hours scouring through the worlds of the other Gods.

Above all, the Church banned any book that mentioned sex, even scientific journals. At the time, sex was not only a word banned in households, it was banned in schoolrooms as well. It was that he was exposed to books which discussed sex at an early age that made possible the most prolific testimony of his ministry,

"We have made of sex the greatest of sins, whereas in itself it is human nature, and not a sin at all."

He took a particular interest in the works of the pioneers of modern day genetics, Darwin and Mendel. He believed that genetic science would one day pave the way to a time when all children would be born healthy.

Above all was Antonio Rosmini's *Origin of the Human Soul*. Rosmini believed that the sole purpose of society was to protect the basic human rights and dignity of every individual; a philosophy chagrin to the Vatican which bestowed basic human rights and dignity on only certain kinds of people. This would become the central focus of his ambitions.

Nevertheless, it was these books together with his father's ongoing guidance which would mold him into the revolutionary he would become. These works would also provide him with the ammunition he needed to have some fun - to play with the minds of his captors at Feltre.

Conditioning children for war

At Feltre, he had his first look at a big city. He developed an interest in chess and was permitted to go to Milan. His anonymous benefactors paid for the trips. Although he was no match for his Russian competitors, in three successive years, he placed third, then second and, finally, first in his class in Western European competition.

Albino made the acquaintance of Russian teens his own age. It was then he first realized it was Vatican propaganda which conditioned its children to hate children of Communist countries. He became aware it was this strategy that enabled a few at the pinnacles of churches and nations to cause the mindless masses at the bottom to sacrifice their lives in war to the benefit of the few at the top. He thought this was wrong.

He found that Russian children were every bit as good as he was. On the other hand, the Russian children found that their western counterparts were not as good as they were. They had heard stories about how Catholic countries treated born-out-of-wedlock children. Yet, they believed it to be Russian propaganda. When they found out it was true, they thought less of their western friends. In their homeland all children of God were seen to be equal. There were no orphans in the streets.

For the most part, for the rest of his days in the minor seminary at Feltre, threatened with excommunication by the masters of Feltre and cautioned by his father, Albino confined his troublemaking to the classroom. He would be an ongoing thorn in the side of his teachers just as his father had been an ongoing thorn in the side of the Church.

The tyrant tortures his masters

One day, Don Filippo read from the Book of Joshua, *"'And the Lord said unto Joshua, See, I have given into thine hand the Promised Land. Take for thyself all the silver, and gold, and vessels of brass and iron, and treasure thereof. Take all ye men of war and go around the cities . . . and ye are to enter and take to the sword all of the non-believers and those who are different thereof, every man and woman and child. Leave not one alive!'*

"Yes," the teacher would tell them, "this was a holy time - the time that gave birth to the Christian dream. Actually, the taking of the Promised Land was also the beginning of Judaism and Islamism. It was a wonderful time. It was the holiest of times."

Albino stopped his teacher, "This was the beginning of Christianity? It was a wonderful time? This doesn't make any sense at all - all of the men, women, children and even infants slain with the edge of the sword? You call this holy? There is something wrong here, very wrong."

Albino reached for his copy of the Old Testament. "This is the second of the thirty-three wars that won for the Israelites the Promised Land - the land that had traditionally been the land of the peaceful Canaanites, *'The city of Ai. For Joshua drew not his hand back, wherewith he stretched out the spear until he destroyed all the inhabitants of Ai . . . And so it was that all that fell that day, man, woman and child were twelve thousand.'*

"It goes on and on, thirty-three cities in all. Hundreds of thousands of men, women and children, I counted them up. You call this unprovoked attack, holy?" He said it in such a way as to question Don Filippo's sanity.

The priest corrected him, "My dear Albino, you don't understand. These Canaanites were evil people. That's why God instructed Joshua and Moses to kill them all."

"You mean," Albino followed, "these people had done the same thing to the Israelites; they had murdered the men, women and children of the Israelites? They were acting in retaliation?"

"No," the priest started to dig a hole for himself, "The Canaanites were a peace-loving people. That is why Moses and Joshua had such an easy time taking their cities from them. They had never thought of the possibility of war as they assumed all people were peace loving people like themselves. They had no defenses other than the walls they had built to protect their children from predatory beasts."

"Then they were clearly wrong. The Israelites must have been a very evil people," Albino shot back.

"Just the opposite," Don Filippo tried to explain, "The Canaanites were the evil people. They were worshipping false gods. Some of them even worshipped idols and statutes. They were all evil, very evil people."

"Oh," Albino exclaimed, "like we worship crucifixes and statues today?"

"No." answered the frustrated priest, "The Canaanites were worshipping images of things of nature like the sun, thunder, and even animals, false gods. We, of course, worship the true God."

"How do we know . . ." Albino stopped himself. He decided not to go down that path. He backed up. "Then they must not have had freedom of religion in those days. I guess those who were most powerful would just kill all the others who did not believe in the God they believed in?"

"Again you don't understand," Don Filippo continued his struggle, "The taking of the Promised Land was the cumulation of all the stories that Moses told. For example, the Canaanites were violating the first four and most important of the commandments that had been taken by Moses from God, Himself, on Mount Sinai: *'I am the Lord thy God and thou shalt not have other gods before thee - Thou shalt have no graven images before thee - Thou shalt not take the name of God in vain - Thou shalt keep the Sabbath."*

"So, that's why Moses put them there," Albino exclaimed. "I always wondered why they were there. After all, if one believes in

God, one believes in God. One doesn't need a commandment to tell one that. I always wondered why those four commandments were so different from the others that tell us right from wrong: Thou shalt not kill, steal and so forth. Also, to me, belief and acceptance of God must come from the heart; only a fool would base his faith on a stone - on a graven image - on a book.

"Yet, nevertheless, it is quite obvious why Moses put them there. He knew belief in many gods and worshipping graven images and using gods' names in vain, and not setting aside a special day for worship, were practices of the Canaanites. Moses knew if he listed them first - God's greatest commandments - his people would have no alternative than to slaughter these 'evil' people and their children. Keep in mind the taking of the Promised Land made them some of the wealthiest men of their time."

Albino stared down his master until Don Filippo pointed him out of the room. In certain cases, Albino got more than detention. His teacher would take the time to write an official reprimand. This happened to be one of them.

To the little boy of Feltre, it was more a case of fun than it was one of sarcasm when he attacked the primitive nature of ritual. Albino drew a comparison between the Christian dancing around his wooden statue of the Blessed Mother at festival time and the American Indian dancing around his totem pole. He pointed out the ritual was the same, just the ideology differed. He compared both of them to their common ancestor, the Cro-Magnon Man, which archeology had recently discovered had danced around an altar of bear skulls tens of thousands of years before.

Being a natural born prankster, he took pleasure in playing with the minds of his masters. He was well positioned to do this. After all, they had only read one book. He had read them all.

The conspirators

In simple matters, he would have one of his friends do the dirty work for him. He would give the idea to one of his classmates and the next day his friend would raise his hand and ask, "Adam and Eve had three sons, Seth, Cain and Abel. After Cain killed Abel, we know from the Bible Cain went on to sire a family. I would suppose then that his wife had to be his mother, Eve, since there were no other women?"

Again, Don Filippo would go for the bait, "Although the Bible does not mention it, it is obvious that Adam and Eve also had daughters." The boy would respond "Then in those days it was okay to have sex with your sister?" Albino, sitting in a corner of the room would have to use all of his willpower to maintain his composure. Nevertheless, they would both get detention.

In the book of Joshua, Joshua makes the sun stand still for a day. Another easy one for Albino's conspirators to handle, after all, it is common knowledge the sun is always standing still; it is the earth that is rotating and moving around the sun.

His friend would raise his hand, "Through the power vested in me by God, I command that the sun stand still not only for a day, but for eternity. So I, too, am a prophet." The boy would tell his teacher. "Now, Father, in that I have made the sun stand still for all time, is it possible that you can make the earth stand still for a day?" The class would laugh. The instructor would try to bluff his way out of it but to no avail. Everyone got detention.

Other fun ones involved Adam and Eve and the long lifespans of the early patriarchs. Albino's best friend Giuseppe would ask, "We know all Bibles are consistent that the approximate time of Adam and Eve was 4000BC. If the Garden of Eden was in the Mid East, how did the Americas get populated?"

Every time the teacher would go for it hook, line and sinker. "Mankind spread to the Americas during the last ice age when an ice bridge was formed between Asia and Alaska."

Giuseppe would respond, "According to our history class, the last ice age took place around 10000BC. How is it possible that God created Adam and Eve six thousand years after the Eskimos walked across the Iberian Peninsular?"

Then Giuseppe would add salt to the wound, "According to the Bible, Noah was born around 2800BC and died around 1850BC. How could it be that he lived almost a thousand years?"

Don Filippo would go for it again, "In those days there was a form of radioactivity that caused men to live long lives."

Giuseppe, "The time that Noah lived was the time of the great pharaohs, the mummies of most of which have been found and are now in museums. Only one of them lived beyond the age of fifty-five. How is it possible that they were not affected by the radiation?" Although Albino would not utter a word or even move an eyelash, Don Filippo knew where it was coming from.

The eyes have it

In more complex cases requiring scientific knowledge, Albino would handle the job himself. One of the first discoveries in genetics was how genes determined eye color. Parents who had brown eyes, for example, would every so often have a child with blues eyes. Although their dominate eye color gene was brown, they also had a recessive blue gene which every now and then, in a geometric pattern, would produce a blue-eyed child.

Albino would ask the teacher, "There are two hundred fifty million Chinese with five hundred million black eyes. There is not a blue eye, nor a brown eye, nor a hazel eye among them. To sire the human race, Adam and Eve would have had to possess all of the color genes of mankind. How is it possible that an entire population on the other side of the world from the Garden of Eden would have only one color gene?"

He would then delve into the details of genetic science and prove the Chinese would have had to have had their own origin. He would back up his supposition with the archeological discovery of the *Peking Man* which had been dated to 1.7 million BC. He would nail the lid closed, the sole surviving primate native to China was the black-eyed gibbon; powerful evidence that man and ape had evolved from a common parent.

He drew a similar comparison of the Australian Aborigine who shared the mottled eye color of the surviving species of gibbon native to Australia. He drew a comparison of Africans and Europeans who share the various eye colors of the surviving primates in their part of the world, the gorilla, the chimpanzee and his dwarfed relative, the bonobo and the baboon. He compared the eye color of the orangutan of Indonesia with the original inhabitants of Indonesia.

He went a step further and drew a parallel between the five surviving human race groups and the other surviving higher primates. He would point out the physical similarities between the Caucasoid and the chimpanzee, the Negroid and the gorilla, the

Aborigine and the gibbon, and so forth; all indigenous to their relative parts of the world.

The boy would educate his peers with what he had learned in his books, "There are fifteen species of higher primates: man, the gorilla, the chimpanzee, the bonobo, the baboon, the orangutan, and nine species of gibbons."

Darwin was unfortunate that the boy Luciani had not come along seventy-five years earlier. Darwin had failed to convince the populace of his theory; the boy Luciani would have succeeded.

The preacher of his time can thank God the boy Luciani did not know what we know today. We now know, as DNA fact, that we are the blood descendents of the Cro-Magnons dated as early as 100000BC. Modern DNA studies are powerful evidence that we do share common ancestors with the other higher primates.

For example, we know today, the chimpanzee and bonobo are much closer in DNA to Caucasoid human beings than they are to any of the other higher primates. This means they are much more our blood brothers than they are related to the other apes of the animal kingdom. They are much more human beings than they are animals; the white man's common brothers in evolution.

Thank heaven for little girls

The following day he began what would become a lifelong struggle for women's rights in the Church.

With his discussion of the higher primates the day before fresh in their minds, he told them, "Of the fifteen species of higher primates, only one recognizes the equality of women; the bonobo commonly referred to as the pigmy chimpanzee. Actually, in the world of the pigmy chimpanzee, woman is dominant; man is subservient. In the world of the pigmy chimpanzee," he told them something that he would tell the world fifty years later, *'God is more our Mother than She is our Father.'* [4]

The Tower of Babel

Don Filippo often had to do homework before challenging the young upstart. A few days later he quoted the *Tower of Babel* miracle which 'created' the world's languages. He told the class, this miracle also covered eye color.

Many historians contend the *Tower of Babel* tale was added after Moses' time. Yet, it is more likely Moses was confronted by the issue himself. One must realize that Moses was two thousand years closer to the time of Adam and Eve than we, today, are to the building of the great pyramids.

In Moses' time, in the tiny geographical area he was aware of, there were more than a dozen different languages - none of them of common derivation. The Egyptian hieratics had been derived from the hieroglyphics or word pictures. The Greeks had a completely different language. They spoke essentially the same language, with some minor variation in dialect, as they did at the time of the building of the pyramids. Actually, the Greeks speak the same language today that they spoke at the time of Adam and Eve. Then there were the Hebrews who wrote backwards and the Nubians to the south who spoke in a half dozen different dialects.

It made sense to Moses' followers that if mankind had a common origin, there would be a common language - at the very least a common derivation of languages. To get around this, Moses made up the *Tower of Babel* story: God created population groups of different tongues and scattered them to the ends of the earth.

One must remember in those days, which is still true today, dialogue in a seminary classroom was entirely one-way from the teacher to the student. Candidates are already drugged with the *Opium of the Masses* before they enter seminaries, otherwise they would never give up sex. By their very nature they accept everything they hear. The very fact that Albino would ignite discussions of various topics was something that his teachers had never run into before. They had no experience how to handle questions. After all, they knew nothing more than was in one book.

---

[1] pg 25 Even today, according to *Canon Law*, an illegitimate boy cannot become a priest. Most of these other requirements have been put aside in recent years by public pressure.

[2] pg 26 *Parish Bulletin Feltre*, 10 May 26. Also, *London Times* 22 Jun 26

[3] pg 29 *Povera Tigre Belluno* 12 Dec 26

[4] pg 37 The population is so convinced God has a penis, that whenever the writer follows the word 'God' with the word 'She' or 'Her', he gets an error message. One of the misfortunes of Luciani's premature demise; he would have corrected the *spell check*.

Chapter 3

The Slaughterhouse of the Human Mind

On his first day at Feltre, meat had shown up on his plate. He found he had little trouble trading it with his schoolmates for vegetables. They were of wealthy families and were accustomed to generous portions of meat and hated vegetables.

From his toddler days, His father taught him it was wrong to kill innocent animals for food when nature provided vegetation. He told him compassion for animals is a basic instinct of man which has been stamped out in the western world by the *God of Moses*, Christianity; the *Opium of the Masses*.

His history books told him something more. In prehistoric times, man had been prey to predatory beasts; his was an ongoing struggle for survival. Imitating the predator beasts he killed in his defense, he learned to eat their flesh. Also, he often lived in places of sand and ice where there was sparse or no vegetation.

As modern civilization emerged, no longer dependent on animals for nourishment, people began to think it wrong. Man's basic instinct of compassion began to surface; it was wrong to kill innocent animals to satisfy one's appetite.

The boy found the early Egyptians were not vegetarians because they lived in a desert where there were few animals. They were vegetarians because they thought it wrong to kill defenseless animals. As a matter-of-fact, they revered animals; four of their Gods were of the animal kingdom.

He saw man's natural instinct of compassion for living creatures most clearly in the God Brahma who ruled the eastern world. Unlike the *God of Moses*, who had been created by a single author at a single point of time, the Hindu God was not a figment of one man's imagination. Rather, He had emerged over countless centuries; the product of the minds of many men who shared a common instinct of compassion for all living things.

In Brahma's kingdom in the east, like the Egyptians in the west before Moses' time, all living things were entitled to share in God's providence. Like the Egyptians in the west, the Hindus in the east revered animals; particularly, the cow.

Conversely, the boy Luciani found the *God of Moses* to be a self-centered one. In *Genesis*, God gives man the right to kill innocent animals to satisfy his appetite. *"I give you dominion over the fish in the sea, and over the fowl of the air, and over the animals in the forest that they shall make meat for you."* Moses' God backs it up in *Genesis 9*. He tells Noah *"Every living thing that moveth shall be meat for you."*

He found Moses' disregard for the right to life of animals extended far beyond man's nourishment needs. *Genesis 22, "Abraham built an altar near the bush and laid the wood in a row for a burnt sacrifice; and bound Isaac his son and laid him upon the altar on the wood. Abraham took the knife to slay his son and the angel gave him a young ram; and Abraham slew the ram instead of his son and offered up a burnt offering."* The law of the *God of Moses* decrees it is morally right to kill animals for mere ceremony; not right to kill children of one's own blood.

So important is the requirement of blood sacrifice to his God for salvation, Moses devotes twenty chapters, including the first eleven chapters of his book *Leviticus* to it; the largest volume of scripture concerning any issue in the Bible. Moses establishes the fundamental requirements of the *God of Moses* for salvation: blood sacrifice of living things and self-adoration. *"I am the Lord thy God and thou shall bow down before me and adore me."*

In an endless parade of ghastly and diabolical rituals, which totally eclipses Hollywood's best attempts at monster creations, Moses forces countless numbers of innocent animals to the axe and pours their blood upon his altars to his God.

Moses preys on the young, *Leviticus 9, "Take thee a young calf for a sin offering, a young ram for a burnt offering, a kid of the goats for a sin offering and a lamb for a burnt offering."*

One by one, in a carnival of blood and morbid gore, Moses drags his tiny victims, each one shivering and yelping in fear, to his satanic altars of death. One by one, he slits their throats and opens their innards. In *Leviticus 4,* he orders his Aryan priests, *"Dip thee fingers in the blood and sprinkle the blood seven times before the Lord; pour the blood upon the horns of the altar that it may make sweet incense before the Lord."*

In bloodthirsty fashion and in ghastly detail, Moses orders his priests, blood dripping from their hands, to place the remains set upon his altar of blood. *Leviticus 3, "Place the fat that is the innards, and the two kidneys, and the fat that is the flanks, and the liver, and so forth. Aaron's sons shall burn it on the altar; a burnt sacrifice, which is upon the wood that is upon the fire; a sweet savor unto the Lord."*

A sweet savor unto the *God Satan*, the creation of a grisly excuse of a madman named Moses. Today, his mindless followers teach their children there is something wonderful and holy about this demonic man. They call his thesis, *"The Holy Bible."*

This horrific practice of killing animals to satisfy the satanic lust of the *God of Moses* continued in Jewish custom into Christ's time. In *Luke 2,* Mary and Joseph *"offer a sacrifice according to that which is said in the law of the Lord, a pair of turtledoves."*

It persisted into early Christianity and paganism; animal sacrifices continued into the catacombs and into the post Roman Empire era. The practice was finally brought to an end in the Christian world by Pope Gregory in 604AD.

In his proclamation, Gregory told the world, *"In His great love for our friends in the pastures and the forests, Christ offered Himself in their place, the Lamb of God, His own body and blood under the symbolism of bread and wine."* [1] Thus began the *Holy Sacrifice of the Mass* as we have come to know it today.

Gregory explained his reasoning in a letter addressed to eastern bishops. *"Christ knew men would go on sacrificing animals to the God of Moses to the end of time. We must ask ourselves, 'Why else would Jesus have offered Himself as a sacrifice?' After all, Christ told us many times sacrifice and ritual has nothing to do with salvation. He had left us with His sole instruction, 'Love thy neighbor as thyself.'*

"Yet, Christ knew most of us would be unable to break free of the macabre sacrificial world of Moses so He ingeniously substituted harmless bread and wine in the guise of His own flesh and blood in place of the lamb. He told us it was wrong to kill an innocent animal to satisfy the bloodlust of any God.

"Also, Christ told us, it was wrong to kill living creatures to satisfy man's lust for the taste of blood which Moses built into us. From all that is known of Him, one should not require scripture or doctrine or my voice to tell one that Jesus, Himself, would never harm a defenseless animal to satisfy His selfish lusts; even if it

were to cost Him His eternity. To think otherwise, would be to believe Jesus would take a young animal and slit its throat and drink its blood to satisfy His lust for taste." [1]

Gregory knew if he were to ban meat entirely he would lose his congregation, so hoping successor popes would add other days of the week to his ban and eventually return man to be the vegetarian Christ willed him to be, he decreed Catholics were not to eat meat on Fridays, a practice which carried into the late twentieth century. So strong are the *gospels* that eating meat is wrong, even today, it is a sin for Catholics to eat meat on Fridays unless they get a dispensation from their bishop. In 1983, *Canon 1251, "Abstinence from meat is to be observed on all Fridays . . . unless a solemnity should land on Friday."*

Abraham offers Isaac

Priest's offering to *God of Moses*

In the Catholic world, Christ is reduced to being a mere implement in the ritual. The offering of the *Holy Sacrifice of the Mass* is made to the *God of Moses*. The wonder and adoration of the congregation is focused upwards toward the *God of Moses*. Christ is simply the sacrifice; the offering of bread and wine in place of the ram used by the mythical character Abraham when he is said by Moses to have offered the first *Holy Sacrifice of the Mass* four thousand years ago. The ideology remains the same, only the implements of the sacrifice have changed.

Catholicism remains today deeply steeped in the satanic rituals of the ancient past tempered only by Christ's substitution of the harmless fruit of the vineyard and the wheat of the fields He ingeniously traded for the lives of the creatures He so loved. *"Do this in remembrance of me."* What else could He have meant?

Of all we know of Him, Christ was not a self-serving God. Jesus is explicit in His instruction that participation in sacrificial rituals has nothing to do with salvation. His testimony is as clear as the sky is blue. There is no substitute for Christ's sole requirement for the Kingdom of Heaven, *"Love thy neighbor as thyself."*

In Search of the Vegetarian Jesus

In studying the Old Testament, the boy Luciani found there was not one God, as Moses' Commandments had decreed. There were, as a matter-of-fact, two Gods. There was the God of fear and evil and hatred of Moses and there was the kinder, gentler God; the one who had appeared to Isaiah and those prophets who followed him. It was obvious to the boy, as it would be to anyone with better than a double digit IQ today, that it is a biblical fact it was the *God of Isaiah*, not the *God of Moses*, who was the true Father of Christ.

In its relationship to Christ, the most important book of the Old Testament is not *Genesis* with its mythical story of *Adam and Eve* as most believe it to be, rather it is *Isaiah* as it contains the only prophesy of the coming of Christ; the only verse in the Old Testament which links it to Christ. It is the *God of Isaiah*, not the *God of Moses*, who promises to send His Son to restore eternal life to mankind; the *God of Moses* never even thought of Christ.

Albino went first to the God he knew to be the true Father of Christ. The compassionate *God of Isaiah* corrects His followers who wrongly mistake Him to be the evil *God of Moses*.

Isaiah1. "To what purpose is the multitude of your sacrifices unto me? I am full of burnt offerings of the rams and the fat of the fed beasts; and I delight not in the blood of the bullocks, or of the lambs, or of the goats. Bring no more vain oblations; your savor is an abomination to me." [2]

Isaiah66. "He that killeth an ox or a lamb is as if he slew a man; he that sacrifices a lamb is as if he cut off a dog's neck; he that offers an oblation, as if he offered a swine's blood, as if he blessed an idol." [3]

So important was His instruction that man respect the right to life of all living things, the *God of Isaiah* creates His Son Jesus as a vegetarian; the sacred promise of the *God of Isaiah* in *Isaiah 7*. *"Behold, a virgin shall conceive, and shall bear a son, and shall call his name Immanuel. Butter and honey he shall eat, that he may know to refuse the evil of blood, and choose the good."* [2]

The boy thought of the great theologians and scholars who through the years had professed to be the 'experts' yet had missed this very obvious point. It would be quite clear to a nitwit there were two separate and distinct Gods in the Old Testament; the true God, the *God of Isaiah*; and the *God of Moses*, the *God Satan;* absolute biblical proof of the age-old proverb: *"Beware of Satan; he comes in many disguises."*

Nevertheless, he turned to the New Testament. He suffered a setback. The story of the *Loaves and Fishes* seemed to confirm Christ condoned the eating of fish. Yet, it may have been Christ was tolerating this weakness of man.

He could find no specific evidence in the gospels Christ had, Himself, eaten meat or fish. Yet, he broadened his search for evidence that Christ had refused to eat meat or fish or had instructed others not to eat meat or fish. He found the following testimony in *Luke 22*.

When offered to eat of the Passover feast which according to the *God of Moses* involved the sacrificial killing and eating of a lamb: *Exodus 12, "Every man shall take a lamb, a male of the first year, shall kill it in the evening and strike its blood upon the door posts of the house within which they shall eat it,"* Christ answers his disciples' demand in *Luke 22, "He said unto them, With desire I have desired to eat this Passover. Yet I say unto you, I will not eat thereof . . ."* In the verse, immediately following *Luke 22*, Jesus substitutes His body and blood for that of the lamb in the harmless form of bread and wine.

As a man, Christ expresses this powerful desire or lust of mankind which causes him to enjoy the taste of the blood of innocent animals which has been built into him by Moses' bloodthirsty God. Yet, Christ, as a God of love and compassion, is able to fight off this desire; He shares the natural love of all living things of the nearby Egyptians and the distant Hindus.

Yet, to the young teen Luciani, this was not good enough. There had to be more explicit proof.

Some changes along the way

There was a small room off to one side in the school chapel which served as a museum. In a case was displayed a huge ancient Bible frazzled and browned by time. Over a period of months, he would sneak into the sacristy and fetch the key and open the case. It was painstaking work for the young teen; he had to translate each word from Latin to Italian.

Finally, he found what he was looking for. *John 4. "For His disciples had gone away unto the city to buy meat . . . Then they came out of the city and came unto him, saying, Master, Eat. But Jesus refused to eat meat, saying, I have meat to eat that ye not know of. Jesus said unto them, My meat is to do the will of him who sent me; and to finish His work I say unto you, Take ye not the lifeblood of the bullock. I give thee the herbs of the soil. Look up thy eyes and look in the fields for they are ripe to harvest."* [3]

He returned to his personal copy of the New Testament; the phrase *'Take ye not the lifeblood of the bullock'* was not there. Yet, the surviving text was there; Christ had refused to eat meat and gave his disciples the alternative of vegetation.

He visited the nearby city of Belluno. In its library he found an eastern version of the Bible. The reference to the bullock was there. He realized 'holy' men through the years had changed Christ's instruction to satisfy their own lust for blood of animals.

For the first time, he understood why the Catholic Church discouraged its members from reading the Bible; it didn't want them to know the truth. It didn't want them to know what Christ really had to say in His testimony. It didn't want them to know of the evilness of Moses, the founder of Christianity. After all, it would not be good for business. [4]

Regardless, the boy came to realize Moses' creation of the *Opium of the Masses* was so powerful that it had drugged the western world into thinking it is not wrong, actually holy and often pleasurable, to kill defenseless animals.

The slaughterhouse of the meat rack

Through the centuries, men in the west have murdered countless billions of animals, not only to satisfy their appetites, but for pelts, sheer trophies and for nothing more than to watch them flee in terror; as is the chase of predator hounds in the hunt of the fox, just for the fun of it.

The boy Luciani was fortunate he would not live to see the horrors of today. Driven by greed, modern capitalistic disciples of Moses submit animals to unbearable torture for the entirety of abbreviated lives. They take hundreds of thousands of baby chicks at a time and feed them arsenic laced chicken feed to accelerate their growth and crowd them, one upon another, into closed concentration camps. A few weeks later, their misery is brought to an end with the buzz saw. The God-giving life span of a chicken is fifteen years; man reduces it to scarcely more than a month.

In 1971, Luciani addressed a letter to mankind. *"In my short time, the population has quadrupled and the miracles of modern science have stretched our lives almost twofold. Yet, the selfishness of our species causes other species to forfeit their natural lives. One day we will answer to the God who made them."*

Milking cows have been robbed of their pastures. They spend their entire lives in stalls scarcely large enough to hold them. They stand for hours on end upon open grates which carry their droppings below. The rule of thumb for the grazing of dairy cows is an acre per cow; a significant cost to the dairy farmer. Today, he needs only thirty square feet; he force-feeds them grain in their bins. The sprawling pastures of yesterday are needed to build housing for the exploding human population.

Yet, in the world of cattle, the little 'girl' is better off; her ability to yield milk will allow her to reach adulthood. Only one in a hundred little 'boys' will be so lucky; one bull can service a hundred cows. The other ninety-nine will feel the thud of the *slaughterhouse of the meat* before they reach their first birthday; it makes no economic sense to feed them costly grain beyond their maturation to size. Besides the older they get, the tougher the meat, the lower the price. A male calf which God willed an average life expectancy of twenty years; man has cut short to six months.

Yet, those are the lucky ones. Half of them will not see four months, as laws of many nations require their execution before that time in order that they can be sold at premium prices as calves.

White Light Dark Night

Author's Collection

Three month old calf tagged for *The Slaughterhouse of the Human Mind*

Good Christians think nothing of it. They devour chickens by the billions in fast food restaurants. They take particular delight in veal, calves liver and other calf entrails. They relish calves brains; *sweet breads*. After all, it is written in their book, *"a sweet savor unto the Lord."*

Mindless followers, who go to church and fall on their knees in adoration of this *God of Moses*, this Monster, the holy *God Satan*, who sanctions the ongoing murdering of billions of helpless animals, as the bells in His Christian world toll.

Even the deer in the forest have lost most of their world to man's encroachment; so much so, it is considered humane today to pour bullets into them for the fun of it to save them from slow starvation. The victims of man's unchecked population growth is not limited to starving children in third world countries; it extends to livestock and animals in the wild and all living things.

In July 1974, Luciani spoke to his congregation in Venice, *"Preachers spend millions of dollars they collect for the poor on campaigns to protect the right to life of senseless specks of fertilized eggs. Not a one of them has spoken out against this ongoing Holocaust of agony and pain and death of these living, breathing, feeling, loving creatures of God."* [5]

Today, in the western *Slaughterhouse of the Human Mind*, constitutions of free nations guarantee all citizens the God-given rights of *Life, Liberty* and the *Pursuit of Happiness*. In the world of righteousness and good conscience, the Hindu world, constitutions of free nations guarantee all living creatures the God-given rights of *Life, Liberty* and the *Pursuit of Happiness;* Christ's instruction in His sacred testimony, *Mark 16, "Go forth and teach all creatures, great and small."*

Parental guidance

Albino brought his findings to his father's attention. "How is this possible? How could this have possibly come about? How could men ignore this explicit testimony in the Bible?"

"It would not be good for business," His father told him. "It would not be good for business for the Vatican for the Christian world to know Christ and His Father had decreed to be a mortal sin the murdering of helpless animals to satisfy man's lust for taste.

"If one reads the Bible literally, it is the *God of Isaiah* and not the *God of Moses* who is the Father of Christ. Aside from His prophecy, the *God of Isaiah* shares Christ's love of all living things. Isaiah not only rejects Moses disregard for the lives of animals, he rejects the superiority of man over woman and certain kinds of men over all other kinds of men. In the world of the *God of Isaiah*, all creatures, great and small are equal. His rule of law is spread among all the pages of His book, *'Love thy neighbor as thyself,'* precisely Christ's message in the gospels."

His father explained how this had come about. "When men first preached the gospels they had no customers. Moses, in his creation of his God, had so instilled hatred of others who were different, greed for gold and lust for the blood of animals into men; these things were far too much to give up in this life for what was thought to be a long shot at the next life. It still is today.

"So they changed their story. They substituted the *God of Moses* for the *God of Isaiah* in Christ's testament. They told those who gathered around him, 'You can ignore what Christ had to say concerning these things. You can subordinate women, enslave blacks and spread your hatred of Jews, bastards, faggots and others. You can live in luxury and ignore your neighbor's children on the other side of the world as they starve to death. You can kill the young of the animal world to satisfy your lust for the taste of their blood. All you have to do is pay me to offer up Christ as a sacrifice to the adoration of the *God of Moses*. I will give you eternal life.' Thus grew the largest church in the world."

He told the boy, "Had He not ascended into heaven, Jesus would roll over in His grave at what has become of His half of the world. That the Catholic world has been fooled into thinking one can reap salvation by paying men to dress up in elegant robes and tall hats and jewels and prance about altars of marble and gold drinking 'blood' and offering it up to the *God Satan* of Moses.

"Rome has replaced Christ's requirements for salvation *'Love thy neighbor as thyself' 'Sell all thou hast and give to the poor' 'Take ye not the lifeblood of the bullock'* with PREJUDICE and pomp and ceremony and wealth and majesty and strings of glass beads and the chanting of vain repetitions."

He told him something the boy Luciani would carry with him all the days of his life, "When Christ gave us our sole requirement for salvation *'Love thy neighbor as thyself,'* Jesus meant all our neighbors, *'great and small.'"*

[1] pg 41- 42 Apostolic Library Vatican
[2] pg 43 Because *Isaiah* spoke of an entirely different God than the God Moses spoke to, 'experts', through the centuries, have tried unsuccessfully to link the *God of Isaiah* to the *God of Moses*. Yet, concerning every major issue the two Gods differ. For example, the *God of Moses* treats sex as the greatest of sins; whereas the *God of Isaiah* scarcely mentions it and in no case condemns it. Christ differs from the *God of Moses* on major issues, yet agrees with the *God of Isaiah* on major issues.
[3] pg 45 this verse is consistent with the oldest of the New Testament, British National Library.
[4] pg 45 John Paul II encouraged Catholics to read the Bible but he changed much of its wording
[5] pg 47 *Messaggero Veneto* 14 Jul 74

Chapter 4

The Worst of People

Not all the things Albino did in class were fun and games. Sometimes things got quite serious. There was a young priest, Don Giulio Gaio, who taught a class on catechism. One day, he was asked by one of the students, "Why is it that we don't have a law against atheists and put them all in jail? They are the worst of people. They are the source of all the world's problems."

This was all Albino needed. After all, his father was a nonbeliever. In fact, his father did not believe that Christ had ever lived. What's more, Albino knew his classmates knew this and the student's question was a direct attack on his father.

Keep in mind, although his father did not believe Christ had ever lived, he did believe fervently in the philosophies of Christ as was set forth in the gospels and he also believed that good men, in an attempt to do away with the evilness of the *God of Moses*, had written the gospels.

When the young priest started to side with the student, Albino didn't bother to raise his hand. To him it was a matter of honor - he would stand and fight to the end in defense of his beloved father who had taught him right from wrong. Hollywood could not do justice as to what happened next.

Bringing his fist down on his desk, he stood up, "Instead of throwing stones, let us first define what we are talking about, so that we all know exactly what we are talking about."

Don Gaio looked at Albino as if he didn't know what he was talking about. The young boy answered the priest's unasked question, "We are talking about God. That is what we are talking about. How one defines 'God'. And that is all we are talking about. So let us define the 'God' of religion versus the 'God' of the atheist." Still there was not much more than a dumbfounded look on the faces of both the teacher and the students.

The priest motioned Albino to sit down but the young boy held the floor, "Let us start with what we know. To begin with, we have the sun. We know the sun holds the earth in its orbit and the sun is the source of all life on our planet. The reason it is the source of all life on our planet is because it is the source of all energy in our solar system.

"We also know, as fact today, from Doctor Einstein's work, that energy is the fundamental unit of God's creation. Without energy, without the sun, nothing could begin, nothing could grow, nothing could move, nothing could be." Father Gaio, raising his voice in a nervous twang, told Albino to sit down.

Yet, the boy didn't budge. "So both religion and atheism have a common definition of 'God'. 'God' is the source of all energy. So we all have a common 'God'. But the similarities end there. For religion does not accept the *God of Nature* - the natural order of God's creation - the God of the atheist. This is what defines a non-believer from a believer. It is what defines an atheist versus a Christian, or for that matter, a Jew or a Muslim." Don Gaio again told him to sit down. Again, Albino ignored the priest.

"We as Christians believe in the supernatural. In short, we believe in ghosts." Don Gaio shot a look of insanity at the boy and the class snickered at Albino's ridiculous comment.

"Yes, we Christians believe in ghosts. We believe a ghost, who claimed to be God, appeared to Moses thirty-five hundred years ago. What's more, we believe in scores of other ghosts that appeared to various prophets since. In fact, we still believe in them today as we accept the modern ghosts of Lourdes and Fatima.

"The basic difference between a Christian and an atheist is: the Christian believes in ghosts and the atheist doesn't." Don Gaio was tongue tied. The class took on a serious look of apprehension.

"First, let us consider the beginning of us. We all know, as a matter-of-fact, we are conceived of tiny bits of energy that come together and result in a chemical reaction. That is the will of the *God of Nature* - the natural order of creation - the God of the atheist. The God we know, as a matter-of-absolute-fact, gives us life. It is an absolute fact of nature the 'egg' comes first.

"But, religion does not accept the will of the God we know, as a matter-of-absolute-fact, gives us life. It tells us, 'NO, the 'chicken' came first.' So we have the story of *Adam and Eve*. And we all know, today, as a matter-of-fact, the Indians were running around the Americas thousands of years before God created Adam and Eve in the Garden of Eden. We also know the Cro-Magnons were running around Europe thousands of years before that."

Albino stopped for a time to allow his listeners to catch up. "Then, we have the end of us. We know, as a matter-of-fact, that each one of us will eventually die and return to become a part of the energy force that created us. That is the will of the *God of Nature* - the natural order of God's creation - the God of the atheist. But, again, religion does not accept the will of the God who we know, as a matter-of-absolute-fact, gives us life. It tells us, 'NO, for a few dollars, I will give you eternal life.'

"And how will we live forever? Religion tells us the sun is only the hand of God, that there is a powerful superman sitting on a great throne who controls the sun and has a large book in which he records everything each one of us does from the time we are born until the time we die - the premise of all religions. So, for a few dollars, we will live forever.

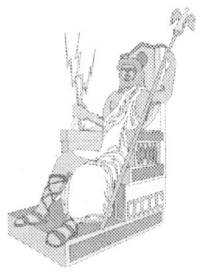   

God of Religion   God of Nature
God of man's imagination   God who gives us life

"Then there is the in-between - life. There are difficulties in life. People suffer from birth defects, diseases, injuries, accidents and so forth. That is the will of the *God of Nature* - the natural order of God's creation. Again, religion does not accept the will of the God who we know, as a matter-of-absolute-fact, gives us life. It says 'NO, for a few dollars I will give you miracles.'

"It tells us to fall down on our knees and pray for miracles - as if to say that God would favor some of Her children over others. To seek miracles is to refuse to accept the will of the God we know, as a matter-of-absolute-fact, gives us life." The young priest tapped a boy on the shoulder and sent him for the headmaster.

Albino didn't blink. "Above all, the *God of Nature* tells us all of Her children are equal. Again, religion does not accept the will of the *God of Nature* - the natural order of creation - the God of the atheist. The God who we know, as a matter-of-absolute-fact, gives us life. It says 'NO'. It tells us women and animals are mere property of men. It tells us of the superiority of the white race and the subordination of black and other races and whosoever be the child that hath a blemish, a blind child, or a lame child, a broken-limbed child, a hunch-backed child, a dwarfed child, a diseased child, a queer child, or a child born out of wedlock is not to approach the altar of the Lord." Pausing, his eyes roamed around the room. His voice took on a tone of bitterness.

"Each one of us here in this room is responsible for each of the lives of those who have gone off the bell tower." He looked at Don Gaio and then at the headmaster who had just come into the room. He raised his voiced just short of a shout, "Why? Why, don't you lock the bell tower?" Then he stared down every student in the class until each one of them looked down away from him.

"Now we know what we are talking about. The atheist believes in the *God of Nature* - the natural order of creation. The God we know, as a matter-of-absolute-fact, gives us life. One who creates all Her children equal. The rest of us believe in a mythical superman who sits on a great throne and keeps track of what we are doing. Who tells us some of us are better than others. Whose emissaries, for a few dollars, promise us miracles and eternal life.

"When one says the atheist does not believe in God, one could not be more wrong. True, he does not believe in organized religion - organized business - but, believe me, he does believe in God. In fact, he is the only one of us who accepts the will of the God who we know, as a matter-of-absolute-fact, gives us life.

"Let us put this thing into definition so that we all know what we are talking about. The atheist accepts the will of the God who gave him life. The Christian does not accept the will of the God who gave him life. He believes he can buy a better life and even eternal life from another man."

Albino stopped as if he had come to the end of what he had to say. Suddenly he started up again. "The day will come when each one of us will have to answer to the *God of Nature* - the natural order of creation - the God of the atheist. The God we know, as a matter-of-absolute-fact, gives us life. We will have to answer to Her as to why we have not carried out Her instructions."

For the first time, in Albino's dissertation, Don Gaio saw his chance to capitalize on the opportunity and attack the boy's logic. Holding up his hand as to stop the boy, he repeated Albino's last words in a question, "Her instructions? Could you be so good as to tell us in what book we might find Her instructions?" A snicker ran about the classroom. The headmaster chuckled aloud.

Albino took his time. He waited for the noise to subside. He told his teacher, "Her instructions are contained in the *Bible of the Atheist* - it is what tells him right from wrong." Don Gaio, this time, thinking he had backed the boy into a corner, licked his lips and smiled to himself, "Yes, I know. But, can you tell me, young man, in which library might I find this book?" This time the class rang out in laughter and the headmaster roared.

Albino stood there quietly with a look as if he had nothing to say. Once again, he waited for the clamor to die down. Then, he told Don Gaio, "Her book can be found in the *Library of the Atheist*." The attention of the class quickly switched back to Don Gaio in anticipation of how he would respond. The teacher decided to go in for the kill. "Yes, my dear boy. But, where is this library? On which street corner may I find it?"

This time the whole class roared as their heads rolled back to Albino who remained silent as a lamb. He seemed to be lost for words. To the observer it appeared Gaio, though badly wounded, had won the battle. But, actually, Albino was waiting for the class to quiet down. Like a lion in tall grass, he was carefully testing the grass and waiting for the opportune moment to pounce his prey.

He wanted to be certain that each and every one of them heard exactly what he was about to say. What's more, he wanted to make certain that each and every one of them would remember it for all the days of their lives. His eyes roamed around the room catching each and every student. They paused briefly on Giulio - the student who had raised the question in the first place, and finally came to rest on both Gaio and the headmaster who were standing together in pending triumph.

"This is the *Library of the Atheist*. This is where he finds the book that tells him right from wrong. Her instructions - the instructions of the God who gives us life - are here." Slowly at first and then more rapidly, he lifted his hand and pointed to his temple.

The Bishop of Vittorio Veneto once told me it was this attack on his father which would make him into the man he would become. Whereas most of his classroom tactics were a matter of fun, in this case, he rose to the occasion. He took on the garb of a great trial lawyer in defense of his beloved father. He would develop, in these few minutes, the argumentative skills, strategic talent and expression which one day would guide him through the courts and legislative forums of Italy in what would be his lifelong struggle for equal human rights and dignity for oppressed peoples.

Despite the fact his nickname 'Piccolo' - little one - would remain with him the rest of his life, regular meals and sports, mainly soccer, had given him an athlete's body of envious proportions. After the class, he told Giulio, "Don't ever try that again. The next time I will use my fists instead of my eloquence."

To all that history has recorded, Giulio never tried it again. Neither did any of Albino's other classmates. Perhaps, not because they feared so much his fists, as they feared his eloquence.

So it went on for five years, the young teenager literally torturing his masters day after day. Each one of them wanted him out of the school as quickly as possible in order that they could survive with their sanity. In that everything he had to say was wrong to them, none of them could rightfully give him passing grades. Yet, it was the only way they could get rid of him. For the money was far too much for the diocese to give up.

So it was on a sunshine afternoon in the early summer of nineteen hundred twenty-eight, Albino Luciani graduated at the bottom of his class in the courtyard at the foot of the bell tower in Feltre. The class valedictorian gave his promise to serve God and the wreaths were passed out: catechism, theology, liturgy and so forth. One by one, his classmates strolled to the platform to pick them up. Had there been wreaths for science, physics, chemistry, archeology, anthropology, sociology, psychology, mathematics, genetics, language, ancient history, world politics, compassion, courage and change - he would have taken them all.

The next day, the boy Luciani was gone. The masters of Feltre would celebrate. But, they would never forget him. No matter how hard they tried, they would never forget him. And they never locked the bell tower, either.

Chapter 5

The Major Seminary at Belluno

In the autumn following Feltre, a substantial grant surfaced from the anonymous donor and the youth Luciani showed up at the major seminary in Belluno. There was no easy way out at Belluno for those who found themselves to be in violation of Moses' laws. No bell tower. One had to resort to a straight razor, an overdose of pills or the combination of a rope and a chair.

Unlike the villages of Canale d'Agordo and Feltre where he had grown up which had an occasional spot of color here and there, Belluno was a sprawling city of sameness. Every house, every building was of the very same shade of beige stucco topped off with orange tile roofs. Even the seminary was of the same beige stucco crowned in orange.

The seminary at Belluno was as large as the one at Feltre had been small. There were as many teachers here as there had been students at the minor seminary where he had *groomed his mane.* At Belluno, Albino was determined not to repeat the record he had at Feltre which had brought him poor grades and close calls with expulsion. This time he would give them what they wanted. He took the early lead, and the others would never catch up. Although deeply immersed in his studies, he could not ignore what was going on in the outside world around him.

The embers of war

One might ask why Italy was Hitler's ally in the war. Being Christian, Italians believed fervently in the superiority of the Aryan race. It was for this reason blacks were not allowed in Italy or for that matter other heavily populated Catholic countries of Europe; the reason why few of them died in concentration camps.

There remained the problem of the Jews who were believed to be of the Aryan race. In 1937, Pius XI commissioned Hitler to conduct the largest investigation ever undertaken to search out the origin of the Aryan race. The objective was to determine that the Jews were not of the Aryan race. Hitler financed the expedition and employed half of his SS guard to conduct the search which spanned half the globe, but no reasonable conclusion was ever reached. Nevertheless, Hitler hated the Jews so much that even though he could not prove they were not of the Aryan race, he tried to annihilate them anyway.

Nevertheless, at the time, Pius, Mussolini and Hitler shared the same ideology, the Aryan race ideology. As a matter-of-fact, Hitler could not have succeeded without the other two. It was the triad of these three men who were the architects of the *Fascist* movement in Europe which eventually culminated in World War II and cost fifty million lives.

In Italy's case, uppermost was the black-white issue. In 1935, Italy invaded Ethiopia, a country made up of fifty million blacks. Mussolini, realizing Italians stationed in Ethiopia might integrate themselves with blacks, imposed heavy penalties for sexual intercourse between Italians and the natives. The minimum penalty for an Italian was five years in prison and for the Ethiopian the penalty could range all the way up to death.

In any event, the pregnancy, no matter how progressed, was most always aborted and any offspring usually 'disappeared'. The objective, driven by the Vatican, was to maintain the purity of the Aryan race. Although it was mostly kept under the table, Italy's occupation of Ethiopia surpassed the Holocaust in infanticide.

Although the United States was no angel itself concerning the treatment of Negroes, the atrocities were so great in Ethiopia that it joined together with its allies and placed sanctions on both Italy and Ethiopia. It was these sanctions which remained in place at the start of the war that made Italy the great ally of Germany.

Fascism had been implanted in Italy several years earlier. When one plans to take over the world at sometime in the future, one starts with the young who will carry out the plan in the future.

The role of the scouts

The role of the scouts in prewar Italy and Germany was precisely what it was at that time and remains today in America, to indoctrinate the young into *Fascism*; hatred of those who do not comply with Moses' laws.

The American Scouts [1]

The *Christian Scouts* grew up out of Mormon Sunday schools in the latter part of the nineteenth century. Early in the twentieth century when the scouts were nationally organized, its founders discarded the name *Christian Scouts* in favor of the name *Boy Scouts of America*. The objective of this name change was to entrap Jewish and other non-Christian children into Christianity, what was commonly known as *Fascism* at the time.

Most racial and ethnic slurs like spic, kike, spook, nigger, faggot, chink, retard and bastard were perpetuated in the scouts; scoutmasters often used these slurs to discipline scouts when they got out of hand by referring to them as these 'freaks' of society. [2]

Most of these terms originated in the scouts whereas others were carryovers from former times. The word 'faggot' originated in France and refers to the charred remains of homosexuals who were burned at the stake. The word 'nigger', to the best that history has recorded, first appeared in the latter part of the seventeenth century. The culprit was alleged to have been a Christian preacher in Charlestown who used the term in referring to Negroes being unloaded from vessels as being animal dung.

Photographer unknown

Boy Scouts of America Wichita Kansas 1920

Be Prepared

"I will do my best to do my duty to God and my country"

White Light Dark Night

God Hates Fags

Today, in the United States, cub/boy/girl scout organizations are active in homophobic activities which include picketing funerals of people who were known to be gay. The scouts wear *God Hates Fags* tee shirts and carry signs and shout obscenities. In 2006 alone, they picketed more than 130 funerals nationwide, including the funerals of three dozen soldiers who had died in Iraq. Search Google: God Hates Fags photos.

In 2000, the Supreme Court ruled American Boy and Girl Scout organizations could bar gay youth from their membership. In another ruling, the same court ruled that scout organizations could exclude children who don't believe in ghosts or the supernatural, that is, so-called atheist and agnostic children. Search Google: atheist boy scout Supreme Court.

The Italian Scouts

For the first quarter of the twentieth century, the Catholic Church used the *Catholic Scouts of Italy* to indoctrinate the youth of Italy into its *Fascist* fold. In 1925, Pope Pius XI and Mussolini began graduating Italy's youth, both girls and boys, from the scout organizations into a new *Fascist* organization. These would be the same boys and girls who fifteen years later would provide Hitler and Mussolini with their military power in their struggle to establish the superiority of the Aryan race.

In April of 1926, the *Fascist* youth movement was given official legal status in Italy and in January of the following year a law was passed in Italy which forbade the formation of any new youth movements and dissolved all branches of the *Catholic Scouts*. That Mussolini and the Pope were grooming these children for war was quite visible; they dispensed with the almost angelic dress of the *Catholic* Scouts in favor of military uniforms for the newly formed *Fascist Scouts*.

In December of the same year, the Pope issued an order to all elementary schools in Italy to enroll all children and teens in the new *Fascist Youth Organization;* as already mentioned, this action by Pius XI prompted thirteen year old Albino Luciani to make his debut onto the public stage.

The Nazi Scouts

By 1933, the process had been completed and Italy's youth had been entirely indoctrinated into *Fascism*, synonymous with the superiority of the Aryan race. However, less than 100,000 had been indoctrinated into its Germany counterpart, the *Hitler Youth*.

Hitler appealed to Germany's evangelical scouts who at the time numbered two million. They were natural prey for Hitler as like the *Born Again Christians* of today, they believed literally in the *God of Moses*. By 1938 Germany's youth had been enrolled in the *Hitler Youth*. By 1938, they were preparing for war.

Joseph Ratzinger (Benedict XVI) and Hitler

Blood and Honor
"I swear by God to do my duty to my country"

The motto of the *Hitler Youth, Blood and Honor,* captioned its mission: to establish the superiority of Aryan bloodlines which was consistent with the *God of Moses'* command in the Old Testament, *"...solely the sons and the sons of Aaron, are to serve in my congregation; others are to be subordinated, cast into slavery or annihilated."* The sacred testament of the *God of Moses* that enabled Hitler to convince his mindless followers, *"I believe, today, my conduct in proposing these things is in accordance with the instruction of the Almighty Creator."*

The oath of the *Hitler Youth* program was drafted by Hitler himself, *"I swear by God to do my duty to my country."* He obviously plagiarized it from the *Boy Scouts of America*, *"I will do my best to do my duty to God and my country."*

Nevertheless, it strikes at the common enemy of both Nazi and American scout organizations: atheist children. It certainly would not be in their best interests to have an army of *Tyrants of Feltre* infiltrating their ranks speaking the truth; organizations that prefer to bring up children in the make-believe world of Moses.

When one thinks of the Holocaust, one tends to think of Jews. Yet, a quarter-million gypsy atheists died in concentration camps. Actually, at the time, Jews were considered atheist in the Christian world in which they lived.

Today the *American Boy and Girl Scouts* treat children who don't believe in ghosts and the supernatural and even gay children as outcasts of society; many whose fathers fought and died in battle for freedom. Deeply steeped in the Old Testament, the *American Boy and Girl Scouts* keep Hitler's dream alive.

Fascism   Christianity

In February of 1929, Pius and Mussolini entered into the *Lateran Treaty*, an agreement that formally united the Catholic Church with the Italian *Fascist* movement. There began a campaign by the Church to preach *Fascism* openly from the pulpit and the following year a special election was held in Italy that invited the populace to vote 'Yes' or 'No' to the question of whether it supported the new *Fascist* government.

Pius XI issued a letter to Catholics in Italy, who comprised 99.9% of its population. It was read from every pulpit throughout the country and published in every newspaper telling them to vote 'Yes' on the issue. The voters went to the polling stations on March 24, 1929. The score was 8,519,539 'Yes' and 155,761 'No'.

This demonstrates conclusively the power the infallibility of the Pope had upon the Italian population at the time. Although he would never waiver from the mission his people had given him, they would string Mussolini up in Milan just fifteen years later.

Nazism   Christianity

In his acceptance speech, as Chancellor of Germany, on February 1, 1933, Hitler issued his New World Proclamation: *"The National Government must preserve and defend those Christian and Fascist principles upon which our nation has been built and which define our morality and our family values"*

A few months later Hitler drafted his *Enabling Act* designed to make him dictator of Germany. The problem he faced was that the act required a two-thirds majority and his Nazi Party was a minority party. Like Mussolini before him, he appealed to the Vatican. In July 1933, Pope Pius XI entered into the *German Concordat* with Hitler completing the union of the Roman Catholic Church with the Nazi movement in Europe.

On August 24, 1933, Hitler held a huge rally in Neukolln Stadium in Berlin. Joining him on the podium was Eugenio Pacelli, Secretary of State and Vatican Nuncio to Germany. Three hundred priests and bishops lined the outer wall of the stadium.

Catholic bishops and priests salute Hitler 24 August 1933

In January 1934, Pius XI issued a letter to the two Catholic parties of Germany - the Catholic Party and the Central Party - instructing them to vote 'Yes' for the *Enabling Act*. The next week Hitler became dictator of Germany. The German government changed from a democracy to a dictatorship. The Vatican was the first foreign state to recognize the new government.

One is on solid ground when one says that the success of the *Fascist* movement in Europe depended entirely upon papal decisions, for neither the Italian nor German referendums would have passed without Vatican support as it was Pius XI and not Hitler or Mussolini who controlled the votes. It was clearly the Pontiff who was the puppeteer who held the strings of fate that would eventually cost fifty million lives.

The first concentration camp

Just a month after Hitler became Chancellor of Germany, on March 12, 1933, he opened the first concentration camp at Oranienburg, a few miles from Berlin. Hitler, himself, did not attend its opening ceremonies. His Vice Chancellor, Franz von Papen did the honors. Joining him were a number of Catholic bishops and priests including the Vatican Ambassador to Germany, Eugenio Pacelli. As Pacelli left the gates of the prison, he smiled and saluted to a group of bishops and Nazi officers who had assembled there. In 1939, Eugenio Pacelli became Pope Pius XII.

Opening ceremonies first concentration camp.

Unlike one might believe, the first inmates of concentration camps were not Jews. They were social revolutionary activists, union leaders, gypsy atheists and gay youths who had been rounded up in gay bars and other gay social establishments.

Inmates were branded with serial numbers and forced to wear colored patches which identified them by category: red for political dissidents - violet for anti-Christians and gypsy atheists - black for social revolutionary activists - pink for homosexuals. By the end of 1934, 150,000 were incarcerated in a network of camps. All but a handful of these would be dead when the first volleys of World War II were fired a few years later; mostly of malnutrition.

Flanked by Nazi leaders, Eugenio Pacelli, Pope Pius XII, signs the German *Concordat* 20 July 1933

The 'handbook' for the *Fascist* movement quickly became Hitler's best seller, *Mein Kampf*. The common friend of Hitler, Mussolini and the Vatican was *Fascism*. Their common enemy was *Communism*. It was their common enemy because it fostered the equality of all men and women. In addition, it was particularly the enemy of the Vatican because it promoted freedom of religion; that is, individuals would be free to practice their beliefs in God privately in their homes and churches and the majority would not be able to use its political strength to force its beliefs on minorities. One must realize that at the time the Catholic Church had the vote.

The spread of Christianity and Fascism

Luciani knew it had been this same view of Christianity that had inspired the Boxer Rebellion in China at the turn of the twentieth century which attempted to drive Christians out of China. The Chinese had never known racial strife despite the fact three of the world's four major race groups were native to Asia. Until that time, ethnic and racial strife had been limited to the West, to those Christian and Muslim countries which cultures were driven by the *God of Moses* of the Old Testament. The Boxers knew all Christian and Muslim wars had been driven by this same God.

They didn't want this kind of influence in the East. Although China had a long history of wars of aggression it had never thought there to be differences among races and this extended to other people who appeared to be different including various creeds, ethnic groups and even homosexuals. Until Christian preachers started to preach bigotry and hatred of people who appear to be different, the Chinese never knew what the word bigotry meant as all people were considered to be equal, for they were all seen to be children of the same God.

The Boxers, who were Caucasoid [3] themselves, knew that the United States, being a Christian nation, held other than Caucasoid races as subordinate peoples at the time. Those of the black, yellow and red races and other ethnic groups living in the United States were made to live as inferior peoples - they were not permitted in white neighborhoods or schools - they were confined to ghettos - blacks were imprisoned if caught using a 'white' toilet or riding in the fronts of trolleys or even sitting in the ground-floor pews of churches. Many of these peoples were persecuted, tortured and even killed in hate crimes. The Chinese knew it was the white Christian preacher who was inspiring this kind of hatred.

In the Korean War - a decade before the *Equal Rights Amendment* - Radio Hanoi capitalized on America's policy of racial discrimination so much so the Armed Forces reassigned black soldiers to non-combat duty. It did not take long for the American Negro to realize he was fighting for a country that considered him to be a subordinate human being and against the Chinese who had no racial discrimination policies whatsoever.

In addition, Luciani knew that the Civil War in the United States had also had its roots in what Moses had to say. He had learned from his history books that Martin Luther's reformation of the sixteenth century had moved the Protestants toward the *left* away from literal belief in Moses and in 1626 Roger Williams brought the Baptist Church to Rhode Island which doctrine was designed to move them back to literal belief in the *God of Moses.*

His history books told him that in 1841 the Baptist Church split into the Southern Baptist Church and the Northern Baptist Church, the former believing in the Tenth Commandment as had been allegedly handed down from God to Moses on Mount Sinai which protected the right of one man to enslave another, *"Thou shalt not covet (desire or take from) thy neighbor his slaves . . ."* The latter no longer believing in the Tenth Commandment. It had been this event that set the stage for the Civil War. Luciani saw this happening all over again as he witnessed the first glowing embers that would eventually fire World War II.

He quietly states his case

During this trying period, although he was often tempted to do so, he never approached the press. When the first concentration camp opened in March 1933, Albino, still short of his twenty-first birthday, could no longer contain himself. He had to break the ideology of the *God of Moses* which served as the foundation for Hitler's *Mein Kampf.*

He took his case to his local bishop - Giosue Cattarossi. Cattarossi was known to be one of the most open-minded bishops in the Catholic world. He was more than willing to listen to the most brilliant of his students of his prized seminary at Belluno.

Luciani took the floor and voiced his objection to *Fascism* in the very same way he had as a teenager when he had questioned the validity of the Old Testament. Referring over and over again to Moses as the *"Father of Fascism,"* Luciani went into great detail challenging every material aspect of Moses' testimony.

He went so far as to prove it was an absolute astrological fact the story of creation was a fairytale, it was an absolute archeological and genetic fact the story of *Adam and Eve* was a fairytale, it was an absolute historical and metaphysical fact the story of Noah's Ark was a fairytale, it was an absolute archeological and biological fact of the species the long life spans of the early patriarchs was a fairytale, it was an absolute archeological fact the story of the taking of the *Commandments* by Moses on Mount Sinai was a fairytale. He proved by in-depth analyses that the *Tablets* had been written by a man of less than ordinary intelligence and organization - the reason why modern Christians have put half of them aside today.

Finally, he attacked the story of the taking of the Promised Land itself as being the reason Moses had told all of his other stories. He told Cattarossi, *"The last one in line when the brains were passed out could easily understand why Moses told his many stories; to convince future generations God had given, what is the most prized land in the Arab world, to the non-Arab Jews."*

He pleaded with the bishop to allow him to take his case public. Cattarossi, convinced by his presentation, with much of his own faith eroded, took the case to his senior, the Archbishop of Venice. That was as far as it would go.

Albino knew that to go to the press would be a strategic mistake. He would not only hinder his chances of rising in the Church; he would most likely be thrown out of the Church. Besides, no one would believe him, for the population was too caught up in the *Opium of the Masses* to know right from wrong. He turned back to his studies and wrote his first thesis.

The Anatomy of Sin

Luciani was an analytical genius; to him analysis was his way of life. He set out to determine what the Bible really had to say about the world we live in. His first target was sin.

In his thesis, *The Anatomy of Sin*, he classifies the relative severity of the various kinds of sin in the Bible as follows.

Gravity of Sin	punishment required by the Bible
Venial Sin	misdemeanor or fine
Forgivable Mortal Sin	death

Unforgivable Mortal Sin   permanent exclusion from heaven

He didn't have to analyze the Bible to determine what it declared was the greatest of sins. It is obvious today, even more obvious then, to the least of dimwits that sex was the greatest of sins. Sex permeates the Bible, particularly the Old Testament.

The first book of the Bible, *Genesis*, deals almost entirely with the evilness of sex. In *Genesis 1*, Adam and Eve eat the forbidden apple and God creates the shame and evilness of sex, *"They saw they were naked and sewed fig leaves together and made aprons."*

This is quickly followed by the story of the Canaanites of *Sodom and Gomorrah* who believed sex to be good and beautiful and a gift from God; they didn't believe there was anything wrong with it.

On the same day God reduces to brimstone the cities of *Sodom and Gomorrah*, He promises Abraham the thirty-three cities of the Promised Land. In his very first words, Moses establishes the purpose of his thesis, *The Five Books of Moses*, which cumulates in the sixth book of the Bible, *Joshua*, the taking of the Promised Land. He carefully fabricates his stories of *Creation*, the *Israelites four hundred years in Egypt*, the *Exodus*, the taking of the *Ten Commandments* and all the others to leave behind the record that God had given the Promised Land to the Jews; the fundamental reasoning of Albino's plea with Bishop Cattarossi just a few weeks earlier.

Then there is the story of *Noah's Ark;* God destroys the world for *'sins of the flesh;'* every man, woman and child. Moses convinces his followers sex is the greatest of sins. Until the twentieth century, this remained the conviction of the Catholic world; sex is the greatest of sins.

Luciani listed eighty-seven specific condemnations of sex in the Bible; eighty-one involving heterosexual acts; six involving homosexual acts.

Moses condemns homosexuality in *Leviticus 18*, *"Thou shalt not lie with mankind as with womankind; it is an abomination."* Again in *Leviticus 20*, *"If man lie with mankind, as he lieth with a woman, both of them have committed an abomination: they shall surely be put to death."* In the Old Testament, this is the only condemnation calling for the *death* penalty for a homosexual act.

In this same book of Leviticus, there are seventeen condemnations of various heterosexual activities calling for the *death* penalty; one involving homosexuality.

The homophobic preacher capitalizes on this single verse to cause dozens his mindless men women and children to show up at Matthew Shepard's funeral with placards. Hypocritical followers of Christ *"Love Thy Neighbor as Thyself."*

Shepard family photos

Matthew Shepard   Scene of execution   Protesters at Matt's funeral

In October 1998, in Casper Wyoming, a young student was tied by two religious extremist youths to a split-rail fence and tortured for hours because he was gay. A cyclist, who mistook him to be a scarecrow, found him sixteen hours later in freezing temperatures. Matthew's face was caked in blood except where it had been partially cleared by tears.

Of its 22,000 verses, this is the only explicit condemnation of homosexuality in the Bible; the reason why it is the only one the general populace is familiar with. Homophobic preachers, with little else to support their hatred of gays, capitalize on it repeating it over and over again to their mindless followers.

There are, however, a few ambiguous verses. There are Paul's opinions in *Corinthians* and *Romans* in the New Testament. One must keep in mind that Paul was not a witness to Christ's testimony. Also, he died in 66AD, four years before the first gospel could have possibly been written.

The only verse in either the Old Testament or the New Testament that provides the ultimate punishment of *permanent exclusion from heaven* for a homosexuality is *Corinthians 6*, *"Be ye not deceived, neither fornicators, nor idolaters, nor adulterers, nor effeminates, not abusers of themselves, nor thieves, nor greedy, nor covetous, not drunkards, nor revilers, nor extortioners, shall inherit the Kingdom of God."*

The word *effeminates*, was used in the King James Bible because its drafters believed gay men were effeminate. Whereas it is true that transgender men are born effeminate, homosexual men are not born effeminate although some of them acquire effeminate traits in this life. In any case, Paul forgot about the girls.

The Catholic Bible uses the wording *boy harlots* in place of *effeminates*. The oldest surviving copy of the New Testament held by the British Library has the wording *sexually immoral*. Through the years, homophobic preachers changed the verse. Nevertheless, Paul decrees these crimes as unforgivable mortal sin punishable by *permanent exclusion from heaven*.

In *Romans 1*, Paul repeats essentially the same list; this time calling for the *death* penalty. Paul's condemnations of sin in *Corinthians* and *Romans* are the most encompassing in all of the Bible. If you are on Paul's list, you get both: the *death* penalty in this life and *permanent exclusion from heaven* in the next life. *Fornication* is sex outside of marriage. If you have ever had sex outside of marriage of any kind you are on his list; Paul gives credence to *'many are called, but few are chosen.'*

Luciani found a few ambiguous condemnations of sex by Christ in the late chapters of the *Gospel of Luke*. Again, he found these did not appear in earlier texts; they had been added by men through the years who thought sex to be evil. In reality, Christ did not think sex to have anything to do with what is right or wrong. Luciani found that rather than condemning them, Christ often defended sinners; Mary Magdalene being one of them.

Christ provides one with the opportunity of forgiveness. He excludes only one sinner from His Kingdom: the HYPOCRITE, the Pharisees who pranced about Herod's Temple of Jerusalem in their rich and elegant attire to be seen and heard of men; forerunners of today's hierarchy in Rome who prance about the Vatican in their rich and elegant attire to be seen and heard of men.

Immersed in treasures of diamond tiaras, ruby and emerald crosses and rings, altars of marble and vessels of gold and priceless art while children all over the world starve to death, they profess to be Vicars of Christ on earth. The HYPOCRITES Christ so often warned of in His ministry.

Luciani concluded, if one accepts Paul's testimony as it has survived today, Christ, like His Father before Him, did hate fags, but He hated a lot of other people too. On the other hand, if one accepts Christ's testimony as it was originally recorded, Christ loves all of his children with only one exception, the HYPOCRITE of His day and the HYPOCRITE of today.

Murder vs. Sex

Nevertheless, that Paul's all encompassing condemnation excludes the most reprehensible crime of murder, attests to either a warped mind or extremely low level of intelligence.

Luciani numbered twelve denunciations of murder in the Bible mostly in the Old Testament; Moses' *"eye for an eye"* and *"the Lord set a mark upon Cain; less any finding him shall kill him"* and a few others. Whereas eight of these call for the *death* penalty, none of them call for *permanent exclusion from heaven*.

Of the eighty-seven explicit condemnations of sex in the Bible, thirty-seven require the *death* penalty and eighteen call for the ultimate punishment of *permanent exclusion from heaven*.

He concluded, according to scripture, sex is the greatest of sins; the reason why 'immorality' and 'sex' are synonymous in the Roman Catholic world today; the reason why there is nothing wrong with living in the lap of luxury enjoying filet mignon and expensive wines while children stave to death. On the other hand, according to the Bible and *Canon Law* today, a teenager forfeits the Kingdom of Heaven for the evil sin of masturbation.

". . . faith and the conviction with which we speak."

A few months after his unsuccessful try with Bishop Cattarossi, Luciani spoke at length of humility at his graduation ceremony. He reminded his classmates to never forget they are mere men. As usual, he spoke slowly and decisively.

". . . The preacher is unique among entrepreneurs - for all businessmen know immensely more about their products and services than do their customers - the reason they are able to sell their wares. But we as preachers know no more about the existence

of a God, or for that matter which God is the true God, than do our customers. As a matter-of-fact, we will know nothing about the possibility of an afterlife until after we are dead. And then, perhaps, we may never know. All we have is our faith - and the conviction with which we speak. We came here with one of these - and the good Fathers gave us the other.

"Now, let us take them with us - let us feed them - nourish them - train them - cherish them - and protect them. For these are the horses of the carriage that will one day take us to our destiny. Let us never use them - our faith and the conviction with which we speak - to spread hatred of any of God's children no matter how different they may live their lives. Rather, let us use them to help bring about a day when all men and women - no matter how scorned by scripture - will be accepted with equal human dignity under the laws of nations. . ."[4]

On July 7, 1935, he was ordained a priest. He had learned to heed his father's words, *"Play your cards carefully, until at the helm of their ranks you will bring change to the Church."*

Albino Luciani – White Light

[1] pg 58 the Supreme Court has ruled that the American scout organizations do have the right to discriminate on the basis of sexual orientation and religious convictions. The author concurs, that as private organizations, they have the right to do so. What the author objects to is word 'American' in the title. America is not synonymous with discrimination. It is the exact opposite. What they died for on the great battlefields of war was *"Liberty and Justice for All."* The American scout organizations should be renamed the *'Religious'*, *'Discriminatory'* or *'Fascist'* organizations they are. The Supreme Court, particularly in that one is speaking of discrimination practices against children, should order an appropriate name change.

[2] pg 58 See *History of Boy Scouts of America* and *Christian Scouts of America*

[3] pg 65 DNA classifies the mass of the Chinese population as Caucasoid. Mongoloids are found in the north and Aborigine in island nations of Asia.

[4] pg 72 *Il Corriere delle Alpi* 12 Jun 35

Chapter 6

His Philosophies and Goal in Life

"I love Christ but I hate Christians." Gandhi said it and Albino Luciani repeated it in the aftermath of the assassination of India's legend in February 1948 in the Cathedral at Belluno. The young priest told his congregation, *"Christ's greatest enemies are today's Christians who pay men to prance about altars of marble and gold and never give a thought to the fifteen hundred children who starve to death while they celebrate their Holy Sacrifice of the Mass."* Nothing pinpoints more perfectly how the man thought.

Then there is his conversation with a youth organization in Venice on the day Paul VI named him a cardinal,

"Never be afraid to stand up for what is right, whether your adversary be your parent, your teacher, your peer, your politician, your preacher, your constitution, or even your God!" [1]

Luciani pays homage to Lincoln for having had the great courage to have defied the written word of his God, the Tenth Commandment, *"Thou shalt not covet thy neighbor his property, including his house, his wife, his slaves, his ox, his ass."*

Here, he also warns against another commandment, *"Obey thy father and thy mother."* As a child he often wondered why it was there at all. After all, if one wanted to survive in the early years it was not an option. He knew why Moses had put it there; he knew that the hatred of people who appear to be different he had built into his religion could best be perpetuated and passed on from parent to child.

In the same session, Luciani paid homage to the leaders of the women rights movement in America. He told the group, *"The United States Constitution, which was put together entirely by Christian men, was based on this commandment insofar as it declared both Negroes and women to be man's property and therefore not citizens. Although we all recall the horrific record of Negroes in America, most of us today are unaware that women too, were not only subordinated, but many were persecuted until the likes of Elizabeth Cady Stanton and Susan B. Anthony came along in the middle of the nineteenth century.*

"Women, being property, were usually sold by parents for considerable sums and being property themselves they had no right to own property, not even through inheritance. Women were generally perceived as being house servants and in the bedroom they were often sex slaves. A man could legally rape his wife in the eyes of the law.

"When children were born they became the property of the man. In the event of separation or divorce the woman had no rights, not even visitation rights, to the children. And women being property, as decreed by the God of Moses, had no right to higher education prior to the nineteenth century, the reason why men dominate American history."

Later that afternoon, referring respectively to the *God of Moses* of the Old Testament and Christ of the New Testament, he warned, *"Only a fool would accept a God out of fear for fear and evil go hand and hand. Nothing demonstrates this more clearly than does the Tenth Commandment of the God of Moses. Accept a God only out of love for Him. Christ is that God."* [1]

How can one believe in Christ and yet not accept the only known written word of the *God of Moses* in the Old Testament? It certainly made no sense to Luciani that society would reject the only known written word of God as passed on through the years in the *Tablets;* that society would ignore the *God of Moses'* sacred instructions that slavery was to be a way of life and that woman was no more than a piece of man's property. Yet, accept other tales of Moses that had been passed on down through the centuries, subject to change and embellishment to satisfy the self-serving interests of preachers and storytellers along the way for a millennium or so until papyrus had been developed to a point that permitted some of them to be reduced to writing.

Although the Church would want one to believe otherwise, Luciani was a revolutionary activist of the very first rank. On the one hand, as long as doctrine did not treat people unfairly he conformed to it. Yet, on the other hand, whenever doctrine placed undue hardship on the lives of innocent people, he stepped in.

As a common priest in northern Italy, it was Luciani's outspoken stance against several papal decrees that attracted the attention of John XXIII and made him into the protégée of this Pope. Luciani, at one time, possessed an intense fear of public speaking and, as set forth herein, John XXIII tricked him into taking his place on the world stage.

On that particular occasion, as a simple priest, Luciani addressed the Vatican cardinals in the Sistine Chapel. It was way back then that he set the stage for what one day would be his papacy, *"Our great enemy is the bigot. He lives in each and every one of us. His ally is scripture. Our ally is conscience. His ally is Moses. Our ally is Christ . . ."*

These words more than any others describe Luciani as the man he truly was; the way he lived his life, the way he wanted others to live their lives. For his purpose in life above all else was to rid the world of the bigot. As he so profoundly stated on that occasion, *"As long as he exists no man is truly free. Not even he, the bigot himself, is free. If one is to be successful in bringing down an enemy, one must know who he is, know where he is coming from, know his armament, know his mental embattlement, how he thinks. And believe me, all the bigot has are words, flimsy words - flimsy arrows. Flimsy arrows that he believes will win for him. Flimsy arrows built in an ancient factory."* [2]

Luciani was referring to the testimony of the *God of Moses* which serves as ammunition for the bigot in his war against blacks, women, homosexuals and others. He made it clear to his audience, the majority of which at the time were the very enemies he spoke of, the war he was about to wage would be fought on all fronts.

His writings

Interspersed between the lines of the heavy theology that permeates Luciani's best seller *Illustrissimi* can be found the basic ideology of his ministry. There is his famous letter to that little mischievous boy that the great Master Geppetto had once created,

"Dear Pinocchio,

I was seven years old when I first read your Adventures. I can't tell you how much I liked them. In you, I recognized myself as a boy, and in your surroundings I saw my own.

My dear Pinocchio, there are two famous remarks about the young. I commend the first by Lacordaire, to your attention:

'Have an opinion and assert it!' This is one of reason. It is the lion. It will win for you.

The second is by Clemenceau, and I do not recommend it to you at all, 'He has no ideas of his own, but he defends them with ardor!' This is one of belief. It is the sheep. It will lose for you.

Think of this, as you go through life, as you run through the woods with the Cat and the Fox and the poodle Medoro,

Your magical friend, Piccolo" ³

Luciani spells out the difference between Lacordaire the progressive socialist on the *left,* and Clemenceau the conservative republican on the *right*. His letter to Pinocchio was republished from *Illustrissimi* in most major newspapers of the world when John Paul published it as a cardinal in 1976. Yet, if one were to read any of the biographical briefs released by the Vatican, he is painted to be an ultraconservative. There is good reason for the Church's misrepresentation; had Luciani been a conservative there would be no ecclesiastical motive for his murder.

Great men do not kill great men for personal gain; they kill them to prevent the assimilation of their philosophical ideologies into society. This is fundamental to the question: Did Albino Luciani's struggle for equal human rights for oppressed peoples cost him his life?

One need not go further than to read his letter to his dear friend Pinocchio, one of his most famous, to determine on which side of the aisle he stood. So take your choice, those biographies, published by Rome, which are designed to cover up the motive for murder and depend entirely upon the blind faith or gullibility of the reader, or this rendition which relies entirely upon the facts.

Again, in his *Illustrissimi,* he points again toward the socialist and away from the republican in his expedition beyond the great wall,

Lucien Gregoire

"Dear Casella,

I have had the good fortune to have visited those places which, as we all know, lie beyond the wall. And for each of us, I have found that the Father provides that we will live beyond the wall as we have chosen to live on this side of the wall.

First, I was granted the privilege of seeing Hell. As I peered in through the gates, I saw an immense room with many long tables. On these were so many bowls of cooked rice and gourmet delicacies as one could imagine, properly spiced, aromatic, inviting. The diners were all seated there, filled with hunger, two at each bowl, one facing the other. And then what?

To carry the food to their mouths they had - in oriental fashion - two chopsticks affixed to their hands, but so long that no matter how great their efforts, not a single grain of delicacy could reach their mouths. Although starving, they could not take of these things.

And then, I was able to peer into Heaven. And, here again, I saw a great room with the same tables, same gourmet delicacies, same long chopsticks. But here the people were happy, smiling and quite satisfied. Why?

Because each, having picked up the food with the chopsticks, raised it to the mouth of the companion that sat opposite, and all was right. So my dear Casella, we must learn here, as we make our way toward the great wall, how to use the chopsticks, else we will not know how to use them when we are on the other side of the wall.

*Your magical friend,
Piccolo"* [3]

Again in his *Illustrissimi*, his condemnation of the critics of those who happen to be born different, even going so far as to reach out to those outside the heterosexual mainstream,

"Dear Figaro,

Well then, who and what are you my dear Figaro? A variety of dress? A mixture of feminine and masculine? Of Orient and Occident?

Poor Figaro, against all these nobles with their coats of arms, these bewigged bourgeois, who themselves do every trespass. They are no better, perhaps worse, than you. Barber, marriage broker, adviser of pseudo diplomats, yes, ladies and gentlemen, whatever you like.

They demand that you alone be honest in this world of cheats and rogues. Do not accept what they say, my dear Figaro, for you, too, are a citizen.

But, sadly, perhaps, your only solution is in revolution!

Your magical friend,

Piccolo" [3]

Here, Luciani talks about the scant evidence condemning homosexuality and transsexuality in the scriptures. No mention of either of these in the commandments and only a single explicit condemnation and less than a half dozen ambiguous mentions of them elsewhere in the Bible. This, as compared to more than eighty explicit condemnations of heterosexual sexual activities, many calling for the *death* penalty and *permanent exclusion from heaven.*

His philosophical mentor

As already mentioned, his lifelong philosophical mentor was Antonio Rosmini-Serbati, a monumental progressive who had forty of his propositions for change in the Church condemned by the Vatican in 1887. Rosmini's thesis dictated, *"The state must serve the individual and not the other way around. Society is properly the means; the individual is the end."* Rosmini considered the rights of the individual, no matter how far removed from the mainstream, to be the sacred duty of society. This became the fundamental philosophy of Albino Luciani.

Luciani was also an ardent feminist activist as evidenced by his letters to Carlos Goldini in his book *Humbly Yours*. This was certainly confirmed in his last papal proclamation the day before he died *"God is more our Mother than She is our Father."*

His ecclesiastical goal in life

Although he had great problems with the *God of Moses* of the Old Testament, he did believe fervently in Christ, and it was his *ecclesiastical goal in life* to somehow separate the love that permeates the *gospels* from the hatred that permeates the *Books of Moses,* to somehow separate Christ from Moses.

Despite the fact that it is the explicit testimony of the Bible, he knew it would be a colossal task to convince the Christian world it was the *God of Isaiah,* and not the *God of Moses,* who was the true Father of Christ; for the credibility of the congregation was with the 'experts' who, motivated by greed and prejudice through the centuries, had cleverly conditioned their followers to believe otherwise. Evil men had conditioned the populace to think Moses a good and holy man, whereas in fact he had wreaked havoc on western civilization for three-and-a-half millennia.

As his father once told him, it was that the people of Christ's time believed in the *God of Moses,* the early evangelists had to tie Christ to Moses. Otherwise, they would have never been able to get Christianity off the ground to begin with.

---

[1] pgs 74-75 *Messaggero Mestre* 6 Mar 73
[2] pg 76 *L'Osservatore Romano* 19 Dec 58
[3] pgs 76-79 Reprinted from Luciani's book *Illustrissimi 1976.* Note: many of the letters included in *Illustrissimi* had been previously published by Luciani

Chapter 7

White Light – his ministry

Six months after the chat with the author as related herein, Albino Luciani was named Archbishop of Venice and five years later became a cardinal of the Roman Catholic Church. Still holding to moderate views on most issues, he moved radically to the *left* concerning doctrines which placed undue hardship on the everyday lives of people. He became particularly outspoken on issues concerning abortion, birth control and human sexuality.

Abortion

Luciani proposed radical views concerning abortion and was convinced that removing the stigma associated with out-of-wedlock pregnancies would eliminate what was then the cause of the greatest share of abortions - family embarrassment.

He recalled his childhood in his memoirs, *"I could hear my mama and aunt and sister talking in low tones and every time I entered the room there was a hush-hush of some kind. Then one day my sister took a short holiday. I was told that she had gone to a neighboring village to rest for awhile. But they didn't fool me at all, for I knew exactly what was going on. And as I fell on my knees that night I vowed that I would someday bring an end to it all. And believe me, I will."*

It was for this reason he became a pioneer of the sexual revolution; including his personal contribution of having been the first, in either the public or private forum, to introduce sexual education into schools. His primary objective was to remove what was at the time the major cause of abortions; family disgrace and embarrassment.

Today's youth are unaware that, until the last half of the twentieth century, all Christians believed sex to be sinful and evil; the reason why 'sex' was not a household word when our great grandfathers were growing up.

It was Luciani's influence, more than anything else, that brought public pressure upon Pius XII and forced this Pope to permit some measure of sexual education in Catholic schools on a worldwide basis which practice was almost immediately adopted by public schools.

Luciani's primary objective had been to eliminate what he recognized as the major cause of abortions, family embarrassment. He wanted to bring about a day when sex would be discussed openly between parents and children. Up until his time, sex was something that was not to be talked about within the family. He looked forward to a time when children would be brought up in a world in which *"sex is seen as being good and beautiful but there is a time and place for everything, rather than in one in which sex was seen to be shameful and sinful."* He knew that Christianity's position that sex was sinful caused many young people to grow up in a state of considerable trauma, resulting in guilt complexes which led to less than healthy sex lives and at times even suicide. One knows to a certain extent this is still true today.

He explained to the press, *"My part in this thing one calls sexual education is to bring about a day when the young girl would no longer think she has gotten herself into trouble as the preacher might lead her to believe, but rather she would realize that she had, indeed, gotten herself into paradise."*

Today we know Luciani was right, for very few abortions are now owed to embarrassment. It is much to the credit of Albino Luciani and the many other fathers and mothers of the sexual revolution that millions of children who might have otherwise been aborted now see the light of day.

Unlike other clergy who simply complained about the problem of abortion, he did something about it. Shortly after being named Bishop of Vittorio Veneto he received a letter from a wealthy man who was terminally ill. Lacking heirs, the man offered to leave his entire fortune to build a large church dedicated to Christ the Savior.

Luciani went to visit the man the very next day and asked him to leave the money to build an orphanage rather than a church. The man held his ground and demanded that the great church be built. Having exhausted every possible alternative, Luciani finally reached into his hip pocket and played his trump card, one that he often used to get his way. *"When I was a teenager,"* he told the man, *"my father made me promise that I would live my life in imitation of Christ, and I have kept that solemn promise. Each time that the fork in the road has come up, often only minutes apart, I have asked myself, 'Now, what would Jesus have done in this case?' And I have often pondered the possibility as to how much better the world would be if everyone were to do this."* He then asked the man, *"Now, what do you think Jesus would do in this case?"*

Two weeks later, the man took up a shovel and broke ground for the new orphanage, one that was designed for infants who otherwise would have been aborted. He was buried a month later in a small nearby cemetery. A simple granite marker was placed on his grave by the grateful Bishop of Vittorio Veneto, *"Each day he breathes new life into the world."*

Some years later, shortly after becoming Archbishop of Venice, Luciani addressed his congregation in the great Basilica of San Marco. Looking up toward its immense dome, he told them, *"We must learn to lower our ceiling height to make room for all of Christ's children."* What he meant by this confusing statement is that Mother Church must cease building great edifices to the glory of the *God of Moses* in order to provide the funds necessary to put roofs over the heads of all of Christ's children.

In the twenty years he served as a bishop and as a cardinal, Luciani never built or dedicated a single church, yet he built forty-four orphanages, many of them equipped with schools and clinics. A monk, one of an army of monks and nuns who had spent most of their lives in prayer, once spoke of him, *"He literally pulled us up off of our knees and put us to work, we monks building and maintaining orphanages and serving as youth counselors, and the nuns teaching class and others caring for those children too ill to come to class."* Most of these orphanages still remain today, one built by the dying man which stands at the foot of the mountain below the bishop's castle in Vittorio Veneto.

The bonding glue of unions

While he withheld public comment on homosexuality, he did much to encourage single persons to adopt unparented children. It was his lobbying in Italian Parliament that made it legal for single persons to adopt children in Italy. An opposition member of the assembly challenged his proposal, "But, that would make it legal for homosexuals to adopt children." Luciani responded,

"The desire to parent children is a basic human need . . . Until the day comes that we can guarantee basic human rights and dignity to the tiniest minority, we cannot truthfully call ourselves a democracy." [1]

Yet, his adversary didn't give up, "But homosexuals have a record of splitting up after the 'honeymoon' is over and this would cause children to lose either one or both parents." Luciani closed the gap on his attacker, *"There are two major forces involved in making for long term loving relationships and regardless of what Rome might believe, sex is not one of them. As a matter-of-fact, sex is most often a declining force in many relationships. It often has very little to do with the long term survival of a union. The longevity of a relationship of two people who parent children that is so important to protecting the rights of children until they reach adulthood depends not on sex, but rather on the two major forces that create long term relationships, love and companionship.*

"When one considers the latter, the homosexual has a great advantage. Two people of the same sex who fall in love with each other make much better companions of each other because they are more likely to share common interests. It is for this reason that children parented by homosexual couples are less likely to undergo the trauma of arguments in marriage and of divorce."

His attacker didn't give up. "Nevertheless, homosexuals are pedophiles. This will put children in great danger." *"To begin with,"* Luciani shot back, *"homosexuality has nothing to do with pedophilia; one is sexual orientation and the other is sexual perversion. Yet, in that most cases of pedophilia involve incest, we must consider the question. If our objective is to prevent pedophilia in adoption then the only logical action is to permit only homosexuals to adopt children who are only of the opposite sex. This would reduce incest to zero. This is because the sex of the victim is determined by the sexual orientation of the parent. If we permit heterosexual couples to adopt children, then children of both sexes would be at risk Nevertheless, on the average, homosexual adoptions reduce the risk of incest in half."* [2]

Needless to say Luciani's biting rebuttal silenced his opponents and the measure passed. Within two years, more than a half-million children, who had previously been confined to the streets, were provided loving and economic support by single parents. Some of these were homosexual couples in which case one of the parents had adopted the child, as it remained illegal for two people of the same sex to adopt the same child.

Very little is known of Luciani's involvement with homosexual parents other than a few short notes written in connection with his orphanages, *"We have found that homosexual couples will take handicapped and less than healthy and attractive children. Most importantly, they will take bastards. Heterosexual couples, on the other hand, go for the cutest babies as if they were shopping for a puppy in a pet shop."* [3]

There is another note written in diary format, *"Dear Mama, I have for many years counseled a young couple. They have great sexual attraction for each other, yet, beyond that they have nothing in common. I have yet to be in their presence when they have not been arguing between themselves or yelling at their children. In addition, they both suffer from a serious ongoing drug and alcohol addiction problem for which neither one has ever sought counsel. Both children, having been bombarded during their growing-up years by the incompatibility of their parents, are now confined to institutions. In that I sanctioned this marriage, I must live the rest of my days with this on my conscience.*

"Last week, this couple came to me on a matter of such great urgency that I had to cancel another appointment. They told me of a neighbor - one of the new single parents in Italy – who was a homosexual. As a matter-of-fact, another man has been living with him for many years.

"I have known of this queer relationship for sometime. Both men are contributing members of the community and spend much of their free time helping out in the parish orphanage. Their two beautiful children, a boy and a girl, are the envy of all who are privileged to experience them.

"One night as they were leaving, I noticed tears in their eyes. They told me, it grieves them that they cannot take all of the children home with them.

"Mama, it is this experience, more than any other, that has caused me to understand the qualifications of a good parent. There is something terribly wrong with a society that thinks that one's sex is what makes one a good parent." [4]

Just three months before his death, Pope Paul permitted Cardinal Luciani to address the Vatican cardinals on the possibility the Church might encourage homosexuals to enter into long term loving relationships as they represented the only population group that was large enough and willing to provide economic and emotional support to millions of children who otherwise would be aborted by women too young or too poor to support them.

Luciani argued that the Church's traditional position exiled homosexuals from society, forcing many of them into lives of loneliness and despair. He followed that the Church's position was one of prejudice, as medical science had proved that sexual orientation cannot be changed and the Bible's condemnation of homosexual acts was scant compared to its vast condemnation of heterosexual acts.

At the conclusion of the session, Luciani had been unable to convince no more than a handful of his audience that the matter should even so much as be discussed. He thanked Paul for having given him the opportunity. He then turned to the Vatican cardinals and told them, *"The day is not far off when we will have to answer to these people who through the years have been humiliated, whose rights have been ignored, whose human dignity has been offended, their identity denied and their liberty oppressed. What is more, we will have to answer to the God who created them."* [5]

Strategy of a Strange War [6]

Forty years before the world's psychiatric and medical communities came to the same conclusion, Luciani reasoned that sexual orientation could not be changed by therapy, that the ability to fall in love is a basic instinct.

Yet, as the psychiatric community tells us today, Luciani found that unlike sexual orientation, sexual behavior could be conditioned by therapy or other circumstances. He reasoned that there are two forces that drive a sexual act, love and lust. He knew when two people are in love, love tends to drive the sexual act and that when two people are not in love; lust tends to drive the act.

He understood then what we are coming to know now; a homosexual male, for example, can be conditioned to have sex with a woman only by changing the motivating factor from one of love to one of lust. It is because he felt strongly that God's children not be products of lust; he opposed this type of experimentation.

Luciani's intermediate thesis *Strategy of a Strange War*, written when he was an advanced student in theology at the Gregorian University in Rome, was based on this subject. As a young seminarian in Belluno he had done much work in the local prison and had found that heterosexual men who were confined for long periods of time did engage in homosexual acts. But he also found *"No matter how long the practice went on a heterosexual male could never fall in love with another male, that lust and not love was the driving force behind such behavior, that when a heterosexual male would have a long term intimate relationship with another male in prison he could never be made to fall in love with his partner, that he would never be able to share an intimate kiss with him. Yes, he might grow to like him and even develop great affection for him but he would never be able to fall in love with him."*

Correspondingly, Luciani drew from his experience in the prisons that when one conditions a homosexual male to enter into a heterosexual relationship he can never truly fall in love with his mate of the opposite sex. *"Yes, he might grow to like her, even develop great affection for her, even parent children with her, but he will never be able to fall in love with her."* He concluded *"Children sired in such marriages would be the product of lust and it is important to all men of good conscience that all of God's children be the product of love, not lust."*

He reasoned, *"Our natural instincts cannot be changed by therapy. We are born with two basic instincts, the instinct of survival and its opposite, the instinct of compassion. Unlike all our other impulses only these two are felt as if coming from the heart.*

"The instinct of survival is reserved exclusively for our work for ourselves. It is why Moses' God - our selfish deity - requires in order to be saved we spend our life in adoration of Him. It is also this instinct that makes us think we are superior to others. It drives

selfishness and material gain in most of us, and in some of us it inspires lying, cheating, stealing, raping, molesting, murder and so forth. In its epitome it expresses itself as hatred of others. Above all, it causes us, each time the fork in the road comes up, to ask ourselves, 'Now, what is in this for me?'

"The instinct of compassion, on the other hand, is reserved for our work for others. It is this instinct of Christ - our unselfish deity - that tells us salvation can only be achieved through helping others. It is this instinct that causes us to realize all children of God are equal. Above all, it causes us, each time the fork in the road comes up, to ask ourselves, 'Now, what is in this for others?'

"Affection is the most common expression of the basic instinct of compassion. The instinct that causes us to laugh and cry also comes from this same basic instinct of compassion. We don't teach babies what to laugh at or what to cry about, they are born with this instinct, and all babies will laugh at the same things and cry about the same things. What's more, they will laugh and cry about these same things for all the remaining days of their lives.

Falling in love

"Who one falls in love with is the greatest manifestation of the instinct of compassion. It is best defined as the creation of a perfect balance of mental energy between two people. At its epitome, it totally neutralizes the instinct of survival, that is, either one of the parties would readily give their life for the other without so much as a second thought; so one knows that such a union can only be made by God. Our instinct that controls who we are able to fall in love with will never change. It would be like trying to condition a person to laugh whenever something bad or terrible happens, and cry whenever something wonderful happens.

"To see this more clearly, one can consider one's own experiences. How many times in one's life has one met a person to whom he or she is physically attracted, a person of unusual beauty and personality, yet one finds, time and again, that one cannot fall in love with that person.

"Falling in love is not an accident, but rather it is clearly an act of nature, an act of God. Unlike the preacher might want one *to believe, it is not a learning process, it cannot be a product of therapy; it is a union that can only be made by God."*

Luciani closed his thesis, *"The preacher, so confused by sex himself, believes that whom one has sex with defines one's sexual orientation."* Luciani felt so strongly about his conclusion that in addition to underlining it, he repeated it four times in his thesis, *"It is quite obvious to me that whom one has the natural instinct to fall in love with is what determines sexual orientation. Not whom one has sex with."*[6]

Today we are coming to know what was obvious to Luciani sixty years ago. It is obvious to a straight person today that he or she cannot be conditioned to share an intimate kiss with a member of the same sex. It is equally obvious to a gay person today that he or she can never be conditioned to share an intimate kiss with a member of the opposite sex. One can no more make a straight person out of a gay person, than one can make a gay person out of a straight person.

It is for this reason Luciani never placed the persecution of homosexuals issue very high on his agenda, for he felt the issue would eventually resolve itself; true heterosexuals would realize there is no power on earth that can cause them to have an intimate loving relationship with one of the same sex. And likewise, true homosexuals would realize there is no power on earth that could cause them to have an intimate loving relationship with one of the opposite sex.

He steps in

Nevertheless, whenever the Church's policies were harsh or inhumane, he stepped in. During the ten years he was a bishop every hospital in Italy was Catholic. It was in Italy that hospital policies first barred visitation rights to other than family members.

The objective was to keep lifelong partners of homosexuals from being present thereby allowing the priest to demand that the dying partner renounce his or her loved one. "Otherwise," the priest would heartlessly tell the dying partner, "you will certainly go to hell!" The original source of this sick policy was a papal decree issued by Pius IX in the nineteenth century which has been adopted through the years by most Christian churches.

As a bishop, Luciani is known to have interceded on behalf of homosexuals who had shared long term loving relationships, in at least six specific cases that were reported in the press. In these cases, and perhaps several others not reported in the press, he ordered hospitals within his jurisdiction to admit longtime partners of homosexuals into critical care units. Despite the fact that his action defied a papal decree, the Vatican never challenged him probably because of the strong following of his congregation or perhaps it feared reaction from the press, which at the time was a great ally of the popular bishop from Vittorio Veneto.

In the early sixties, when the *Christian right* waged its war against equal rights for blacks, he issued a statement of praise for President Johnson and his followers *". . . from whose courage will rise a Technicolor world of tomorrow from the black and white chaos of yesterday."*

When civil rights legislation was enacted in 1964, the *Christian right* in the United States, having lost its quest to keep the black in his corner, turned its hatred and its persecution efforts toward homosexuals. By the end of the decade almost two million homosexuals had been arrested and incarcerated; those in northern states as a matter of harassment for relatively short sentences and those in southern states as a matter of persecution for long terms. Two states, Louisiana and Alabama, came within one legislative vote of requiring the death penalty for a single homosexual act.

In the spring of 1967, acting on a tip from a neighbor, police broke into the home of Robert Wise and Timothy Wilson, both 22, just outside Augusta Georgia. Armed with cameras, they caught the couple in their bedroom. The two were tried and convicted of committing a homosexual act and were sentenced to twenty years under Georgia law. Timothy Wilson served only four days of his term. He cut his wrists and bled to death in his cell on his twenty-third birthday.

Bishop Luciani addressed his congregation in Vittorio Veneto a week later, *"The United States is a great nation, but it is a young one. This rings of the Inquisitions and of Salem witch hunts and of slavery, things that we have already put behind us. How many more must suffer and die before we too put this one behind us. How many more must suffer and die before men and women of good conscience rise up and say, 'It makes no difference how much* of Moses' evilness has crept into the laws and constitutions of nations. All that counts is what is right and what is wrong.' This action is clearly wrong. It is clearly wrong to any man or woman of good conscience."[7]

Twenty years after the conviction of Wilson and Wise, the United States Supreme Court upheld the constitutionality of the Georgia law. In August of 1987, Chief Justice Rehnquist led the Court in a 5 to 4 decision that upheld the Georgia law as providing fair and equitable justice for homosexual acts committed between consenting adults in private. Justice Stevens, who had led the opposition, spoke of the decision, *"It strikes against everything that is good and decent and mostly it is a strike against the most basic of freedoms, the right to be left alone."*[8]

Timothy Wilson did not stand alone in his demise for during the ensuing twenty years homosexuality surfaced as the leading cause of suicides among children and teenagers in Bible-belt states. In all, tens of thousands of American children and teens born to parents whose minds had been deranged by the hatred of Christian preachers took their own lives.[9]

The boy on the fence

In the spring of 1969, the local newspaper boy delivered a copy of the New York Times to the Bishop of Vittorio Veneto. The bishop opened the paper and his eye caught a headline, "Police Murder Young Gay." Beneath it was a photo, a view looking down toward a dark alleyway. There impaled facedown atop a heavy iron picket fence was a small-framed boy. Although the picture had been taken at night - as it was dark and fuzzy - one could plainly see that four or five of the spikes had completely penetrated his body from his neck to his right thigh and that the tips of the spikes were wet with blood.

As Luciani read the article, he came to realize the caption was wrong. The boy, a thirteen year old Hispanic youth, although in critical condition, was still alive. It took several hours for the fence to be cut with blow torches to free the boy and the young teen was removed to St. Agnes, a Greenwich Village hospital, with the spikes still embedded in his body. The boy had been arrested by an undercover cop in Washington Square Park and brought to the Tenth Precinct Station for booking.

Fearing disclosure to his parents, the youth, sobbing relentlessly, pleaded for them to let him go. The boy asked to use a restroom where two officers were allegedly overheard threatening to force themselves on the youngster and the boy was either thrown from or leaped out of a window into the dark night and landed atop the fence. The officers involved were placed under suspension pending an investigation of the incident and were eventually returned to active duty when a witness who occupied a booth testified that he had seen nothing; he had only overheard the confrontation and the most he could come up with is that one of the officers used the term, *"little faggot."*

Whereas others blamed the police, Luciani sent a telegram that placed the blame elsewhere. *"It is quite obvious the motive of these preachers is not one of morality, as they would want one to believe, it is clearly one of hate and prejudice and in the case of this boy it is murder. Cold blooded murder. It is not so much these men of the American Gestapo that are responsible, although they are the direct instrument of this dreadful deed, but rather it is the preachers and politicians, who in perpetuating this kind of hatred, are the real killers, the real killers of the boy on the fence."*[10]

A few weeks later the boy died and the following week homosexuals in New York City, infuriated by the boy's death, for the first time stood their ground and fought off the police in what became known as *Stonewall*. The Gay Revolution had begun, the prediction Luciani had made in his famous letter to Figaro had become a reality.

Four years later, in December 1973, the vast membership of the American Psychiatric Association unanimously adopted the resolution that homosexuality is a matter of instinct and is not a matter of mental illness. At the same time it ordered its members to begin the work of removing the stigma that had long been associated with it. Conversely, at the same time, it reclassified homophobia from normalcy to abnormal behavior or mental illness.

Luciani, referring to this ruling, got himself into trouble with his congregation when he made the remark, *"I wonder how long it will take for the sheep to understand this one."* He was referring to the fact that even after Galileo proved through his *Falling Bodies Law* that the earth was round and rotating on its axis, the mass of Christianity continued to believe that it was flat until years later when Magellan sailed off into the west and returned from the east. More than half of them even refused to believe it then.

Luciani was right. Today, more than a third of a century has passed and the stigma remains very much alive. Population and clinical studies place the percentage of homosexuals at between five and eight percent. At a given time, there are more than five thousand Hollywood stars in the public eye of which at least three hundred are gay. Yet, only a dozen or so have had the courage to admit it.

There are well over a thousand players in the NFL and only two are known publicly to be homosexuals - the 'fathers' of two beautiful children; valiant champions of the very first rank. The rest of them remain hiding in the 'closet'. They have no fear facing two tons of humanity coming at them on the football field, yet they are too afraid to admit publicly their God-given sexuality.

In the fifties and sixties, born-out-of-wedlock children, or *little bastards* as they were commonly referred to, were a rung below *little fagots*. A wave of celebrities owed up to their unholy origin. Overnight, the stigma disappeared.

Those few celebrities who have had the great courage to have owed up to their sexual orientation are great role models for today's gay teenagers who are growing up in a difficult society. As more and more homosexuals owe up publicly to their orientation, the stigma likewise will disappear.

Society, as it has in so many other cases, blacks, women, chinks, spics and so forth, will put the matter behind it and begin to concentrate on the real problems of the world.

The Christian preacher will turn his hatred toward his next victim, the growing population of atheist children who are being brought up in a world of reality rather than his world of make-believe.

Rumors

It was Luciani's position on homosexuality that gave rise to rumors that he, himself, was a homosexual. Specific allegations arose in 1976 when the French physician-priest, Marc Oraison, made public his homosexuality. Citing the fact that medical science had proved sexual orientation was a matter of natural instinct; Oraison declared that homosexual love was God's will.

In a statement to the press, Luciani warned Oraison, *"If a priest preaches as he does, everything is ruined."*[11] Although confusing on the surface, the remark obviously meant Oraison should have kept his sexual identity private, that one can best help an oppressed group by appearing to be an outsider; one was far less effective if one appeared to be trying to help oneself. Luciani's statement gave rise to a rumor that he himself was keeping his own sexual identity a secret, although this was most likely not the case as if this were true he certainly would have been a fool to have made such a comment publicly. Nevertheless, the rumor persisted.

Some of the tabloids took this a step further and suggested that Luciani was a practicing homosexual. They reported that the cardinal frequented a square that was notorious for homosexual activity. They backed it up with pictures of him walking through the park. The photos were obviously of him performing his morning office in the small plaza *Piazzetta dei Leoncini* that fronted the cardinal's palace in Venice. By coincidence the tabloids were right in that the park was a haven for homosexuals.

Very often, he would try to bring attention to medical science's discovery homosexuality was a matter of instinct, something the Church's clergy chose to ignore. Mimicking them, he would say in sarcasm, *"Psychology, the science that explains human events, virtually excuses homosexuals. The fault lies with parents who didn't discipline their children in God's law?"*

In his book *Illustrissimi,* in similar cynicism, he writes a letter *to Luke*, the only gospel which mentions homosexuality. He tells the evangelist of modern science's discovery. Anything to ignite some level of discussion in the Vatican of the two: *the thesis of man-made scripture* versus *the thesis of scientific fact*.

There was the added fact that he had moved his secretary Lorenzi into his relatively small apartment in the Patriarch's Palace to make room for unwed mothers. Lorenzi was an attractive almost angelic looking young man, enough so as to give the tabloids added fuel for their fire. But because Luciani was so well liked by the legitimate press, it never pursued the allegations and eventually the rumors tapered off.

Regardless, for Luciani to have spoken his piece on the world's stage must have taken immense courage. He never for an instant thought of himself, of what people might think of him, he thought only of others. A couple of days after the Oraison incident he was asked why he helped *"those kinds of people."* Alluding to the quarter-million homosexual teenagers that had been murdered in concentration camps, he replied, *"If we are ever to be truly free, we must stamp out what Hitler stood for, once and for all."* [12]

The world's children

Concerning what he considered was a more important issue, Luciani became particularly outspoken about the population explosion. He argued that the Church's position on birth control was aggravating the problem; the unprecedented population growth rate was creating massive poverty and starvation at its fringes; particularly in the poor countries of India, Latin America and Africa. Then there was the fact that the Church's position on birth control was resulting in untimely pregnancies that forced abortions in the United States and Europe. The Church's policy on birth control was in direct conflict with its policy on abortion; the Church was itself, in many areas of the world, the underlying cause of the lion's share of abortions.

Exactly what changes Luciani would have made had he lived will perhaps never be known. But what one does know as a matter-of-fact is that he died on the very eve of that time that he was about to announce to the world a change in the Church's ban on protected sex, something that horrified many of those cardinals with whom he shared the Vatican.

Luciani's positions on birth control and homosexuality were particularly dangerous to the *Vatican Curia*, that cluster of twenty or so cardinals in Rome which controlled the *right* in the Church and had traditionally held a stranglehold on the papacy. Not so much in that his positions constituted departure from scripture as the instruction of the Bible in these cases was quite flimsy and ambiguous anyway, but rather because they struck at the very survival of the Church as the largest in the world.

The Roman Catholic Church, like most other churches, depends primarily on large families for its membership since it depends largely on the indoctrination of children before they reach the age of reason. Encouraging smaller families would result in a dwindling Catholic population that would eventually reduce its influence in the world.

Planned Parenthood

On April 11, 1970, Luciani addressed a general assembly of the clergy of the Veneto country in Venice. *"It is easy today to find persons who use the pill and other contraceptives and do not believe they are sinning. If this were to happen it would be best not to disturb them. Let us pray that the Lord will help the Pope to resolve the question whether Catholics should be able to use artificial birth control, particularly in its role in preventing poverty and starvation of the world's children. There has never been such a difficult question for the Church particularly in the intrinsic implications as it affects other doctrinal issues."*[13]

By *"other doctrinal issues,"* Luciani was referring to sex in general inside and outside of marriage including homosexuality. If two people could have sex without any intent of having children then why can't all people have sex without intent of having children? This brings us to the most fundamental difference Luciani had with Church doctrine. He did not accept the thesis that sex outside of marriage was sinful. In fact, he did not believe sex in itself was sinful. Also he knew that it was Mother Church's overly obsession with sex which clouded her vision to see the truth, what is really right and what is really wrong.

When he first became Patriarch of Venice, his personal secretary was Father Mario Senigaglia. Senigaglia and Luciani spent much of their free time discussing the moral cases of their parishioners. *"He was a very understanding man,"* recalls Senigaglia, *"Very many times I would hear him say to both married and unmarried couples and to teenagers, 'We have made of sex the greatest of sins, whereas in itself it is human nature, and not a sin at all.'"*[14]

Senigaglia recalled Luciani would repeat this byte to youth groups in what were the first sexual education classes in the western world. *"He would say, 'Always think of sex as being good and beautiful - a great gift from God. But, also keep in mind, as with all other gifts from God, it comes with responsibility, both to yourself and your loved ones.'*

"He would counsel couples contemplating marriage that it is irresponsible not to engage in sex before marriage. 'Sex is a complex issue,' he would tell them. 'Unlike Mother Church might pretend, it is not the whole of marriage. Yet, it is an essential part of marriage. It is prudent to test the waters before one drowns'"

Two people contemplating marriage should be certain they are sexually compatible before they commit to having children. The only way to test compatibility is to experiment with sex before marriage. This required contraception outside of marriage.

As Archbishop of Venice, Luciani was faced by an abortion toll of two million a year in Italy, a country one-fifth the size of the United States. He initiated the concept of *Planned Parenthood*, which has since been referred as the number one enemy of the Roman Catholic Church. When word reached Rome that he was preaching this kind of thinking he was summoned to the Vatican and although no one knows what he was told, it can generally be assumed that he was instructed to cease preaching this message from the pulpit.

On his return to Venice, perhaps because he felt so strongly about the issue, he was greeted by tabloid headlines suggesting he was planning a schism. Schism - an incumbent pope's greatest nightmare - occurs when a member of the clergy separates from Rome and takes a large part of the congregation with him.

There is nothing in the legitimate press that says Albino Luciani ever considered a schism. On returning from Rome he ceased to preach *Planned Parenthood* in public but very obviously continued to encourage it among his congregation in private. One knows this because when he completed his reign as Patriarch of Venice it had the lowest birth and abortion rates of any metropolitan area in all of Europe. Today, of course, *Planned Parenthood* is the way of life in all first world countries.

In general, concerning matters of personal intimacy, Luciani could never understand how a group of old men in Rome, who had never been in the bedroom, could take it upon themselves to tell others what they can or cannot do in the bedroom.

Papal election

In 1974, Cardinal Suenens proposed to remove the power of papal elections from the *College of Cardinals* to the *Worldwide Conference of Bishops*. Luciani immediately issued a statement to the press strongly supporting his friend's proposal. This would become one of Luciani's most dangerous positions to the conservative core of cardinals who would one day share the Vatican with him. The closer one gets to the congregation the more liberal is the thinking. Like his good friend Cardinal Suenens, Luciani felt strongly that by bringing the election of successor popes closer to the pastoral level, the Church could more readily respond to the needs of its congregation; something they felt was necessary to bring it into rapid reconciliation with the emerging scientific and technical revolution.

Test-tube babies

In July of 1978, Pope Paul was in seclusion at Castel Gandolfo when most cardinals, in support of the decree of Pius XII prohibiting genetic research of any kind, condemned Louise Brown, the world's first artificially inseminated child, as being *"a child of the Devil."* Luciani reached to the *left* and took up his pen and sent the following letter to the parents of the new born child,

"My very personal congratulations to you on the birth of your little girl. I want you to be assured there is reserved for you and your child a high place in Heaven.

Albino Luciani, Patriarch of Venice"

Concurrent with the sending of this private message, he released the following statement to the Italian press,

"I have sent my most heartfelt congratulations to the English baby girl whose conception took place artificially. As far as her parents are concerned, I (the Church) have no right to condemn them. If they acted with honest intentions and in good faith, they will be deserving of merit before God for what they wanted and asked the doctors to carry out." [15]

The message was viewed as the most defiant rebuttal of a papal decree in modern history by a ranking prelate of the Church.

Rumors surfaced both in Venice and the Vatican that when his reconfirmation came due later that year, it would not be forthcoming. Most people believe a cardinal, once appointed, is forever a cardinal. This is not so. Cardinals are appointed for five-year terms and require the reconfirmation of the reigning pontiff when their terms lapse.

One might wonder why Luciani came out in defense of artificially inseminated children at a time when it would obviously aggravate the possible attainment of one of his greatest objectives - to provide loving and economic support for the then two million orphans in Italy, the reason why he had lobbied in parliament to make it legal for single persons to adopt children, the reason why he had petitioned the Vatican cardinals to encourage long term loving relationships between homosexuals, the reason why he objected to the Church's ban on remarriage. For all of these would provide parents for orphans. In addition, there was the sizable sterile population, also a reservoir of parents for orphans.

Artificial insemination would make it possible for both sterile and homosexual couples to have their own children. This would rob Italy of the two largest population groups that would otherwise be available to solve the nation's growing orphan population. One could say that compassion was a part of his reasoning in his defense of Louise Brown, but it was more likely Luciani's vision for the future.

A week after giving his statement to the press he was challenged by a reporter on this very point; artificial insemination would aggravate the orphan problem in Italy. Luciani responded,

> *"My good friend Einstein once told me he could not accept the existence of God because he could not accept that God would play dice with His children, that there is something horrific about how God goes about making children, that millions of fertilized eggs are cast into the sewer in everyday sexual intercourse, and many others are born severely physically and mentally impaired and some with diseases who are destined to suffer unspeakable lives and die unspeakable deaths.*

"Here on the realization of this great event in genetic research, from a practical point of view, certainly from a humanitarian point of view, it makes no sense for man to allow God to continue to have His way in this thing. I believe that artificial insemination will eventually lead to man's greatest achievement, the creation of a perfectly healthy child every time.

"What is important is not how many babies are born, but what is important is that every single child that is born has an equal chance at a good and healthy life. Genetic research will eventually take us there."[16]

Thirty years before, Albert Einstein had commented on Luciani's work, *"Luciani thinks of things today, as the rest of us will think of them in centuries to come."* [17]

It is that Einstein had such a strong resistance to religion that caused researchers to determine the great majority of macro-geniuses in modern history were in the same boat - they did not accept the existence of a Supreme Being. In 1901, Einstein told a reporter, *"Faith is the great enemy of truth."* Reacting to Einstein's remark, a Vatican cardinal told the same reporter, *"Fortunately, ninety percent of the population has an IQ under ninety-five."* This is still true today, ninety percent of the population has an IQ under ninety-five and ninety percent assume there is a Supreme Being.

Remarriage

A few years before becoming a bishop, he wrote an editorial directed at Rome suggesting both Hitler and Mussolini be excommunicated on a postmortem basis. Pius saw this as an opportunity to silence the young rebel from the foothills of Belluno once and for all. He answered Luciani's editorial in his own comment to the press, *"We must learn to forgive our enemies as Christ has taught us."*

It has been said of Luciani that he was a poor strategist. As a matter-of-fact, on several occasions he said so himself. But in reality he was a brilliant strategist. He had acquired this ability when he had been at Feltre; his passion for chess.

In this case it seemed that the Pope had won the game. Bishop Giuseppe Siri, later to rise to become a cardinal and leader of the conservative *right* in the Church, belittled Luciani in the press,

"This young and inexperienced whippersnapper of a common priest from Belluno is no match for his Holiness. The two men are quite obviously on completely different levels of intellect. Luciani is dealing in matters that are beyond his ability to comprehend."[18]

Contrary to what Bishop Siri may have been trying to suggest, one must keep in mind that during the twenty years he served as a bishop and as a cardinal, Albino Luciani was considered to be one of the most brilliant men in the Church, actually in the history of the Church. John XXIII, many times, referred to him as *"The Einstein of the Roman Catholic Church."*

Of all the persecutions impose by the Church on innocent people, it was the Church's position on remarriage that tormented him most. He just could not accept that the Church could take it upon itself to refuse sanctification of the mental union of two people who, having had made a mistake in choosing a lifelong partner at the age of twenty, at thirty had fallen truly in love. This issue troubled him deeply in that it condemned hundreds of millions of people to live lonely desolate lives who otherwise would have lived out their lives in long term loving relationships.

His first commission

His first post had been in the little town in which he had grown up, Canale d'Agordo. He was a priest for less than a month when he was sent by his rector to visit the rector's sister who was terminally ill. The woman had remarried and her first husband was still alive. He was told he had to convince the woman that she must renounce her husband before her death or she would not be buried in consecrated ground and therefore would be forever barred from entering the Kingdom of Heaven.

When he arrived at the hospital he was assigned still another task. The woman's doctor met him outside of her room and told him that the woman had only a few hours to live and asked him if he would tell her this. One must keep in mind that although he would be dealing with a woman who had remarried, she was, nevertheless, a devout Catholic.

He entered the room and taking up her hand he told her the bad news. He reminded her it was Church doctrine, unless she was to renounce her husband before she died, she would certainly go to hell.

Much to his astonishment she answered him with a question, "What do you think I should do?" Being caught up between doctrine and conscience, he asked, "Do you love him?"

Without hesitation, yet, with great emotion that caused a tear to roll down her cheek she told him, "With all my heart."

Albino, her hand still in his, fell silent for a moment or two wondering what he should do. Suddenly, his eye caught a standup crucifix which sat on the table beside her bed and he had his answer, *"Now, I wonder what Jesus would do in this case?"*

Without any further hesitation, other than to attempt to roll a tear back into the corner of his eye, he told her, "Then cling to it, your love for your husband. Don't ever give it up. Not for your brother, not for Pius XII or for that matter all the popes that have reigned before him and will reign after him. For your love for your husband was not given to you by men, but rather it was given to you by God, and He would not be happy with you if you were to give it back to Him to satisfy the whims of common men.

"I promise you, if you have the courage to do this thing for me, there will be reserved for both you and your husband a high place in heaven. Believe me, if it takes me all the remaining days of my life, I will make this possible for you."

Then reaching over and passing by the crucifix he picked up a small framed picture of her husband. Unwinding rosary beads from her hands, he placed it in her hands. He stayed with her until she died four hours later, still clutching the picture in her hands.

When he returned to the rector's house and told him the news, the rector was livid. He blamed the young priest for having failed in his duty. After calming down, he asked Luciani if at least he would falsely attest to the fact that she had renounced her husband in order that she could be buried in the Church. Albino refused. The rector shortly transferred Albino to another parish.

The woman was buried in a remote cemetery reserved for outcasts. Her brother and her family didn't show up. Only a few loyal friends and her husband were there. And, one more, Albino Luciani who, in defiance of a papal decree, said the prayers and gave the eulogy and consecrated the unconsecrated ground.

Checkmate

So Luciani had a personal motive in what was to be one of his most publicized and certainly his most vicious attacks on the Vatican; his demand Hitler and Mussolini be excommunicated. He answered Pius XII and Bishop Siri with a second editorial in which he challenged the Church's authority to excommunicate and condemn to eternal damnation millions of young people who, having made a single mistake in their initial choice of a lifelong mate, had remarried.

He demanded the authority to grant annulments be removed from Rome to the local bishop level, *"I am greatly tormented that Mother Church would see it as her duty to close the Gates of Heaven forever to so many young innocent people who have at last found true love and yet on the other hand see it as its duty to leave the Gates of Heaven open to the likes of Hitler and Mussolini."*[19] The largest tabloid in Italy, capitalizing on Luciani's fondness for chess, gave the incident front-page coverage with the bold headline "ACACCO-MOTTO!" "CHECKMATE!"

The Vatican refused to comment to the press concerning Albino's response. Instead, he was summoned to Rome. Pius XII intended to excommunicate him but Giovanni Montini - at the time Undersecretary of State and later to become Paul VI - stopped the action. He warned Pius such a move would cause an uproar in the press which had surfaced as a great ally of the young priest.

Montini was more than familiar with the young priest from Belluno. In September 1943, Luciani had approached Montini to use his influence to gain asylum in the Vatican for five hundred Jews including two hundred seventy-five children who had shown up on a boat in nearby Naples. Montini struggled with Pius for a compassionate decision. Instead, Pius ordered the boat dispatched to Germany where they died in concentration camps.

Yet, these events marked a critical turning point in Albino Luciani's career; they had won for him the favor of Giovanni Montini who would eventually rise to the papacy - Paul VI.

Nevertheless, Albino was chastised by the Pope. He was told that aside from insulting the Vicar of Christ on earth his action bordered on heresy and was issued a letter requiring him to clear all future comments he had concerning doctrine with the Vatican before going to the press. The letter was structured in a way that if he were to violate its provisions he would automatically self-excommunicate himself. Luciani did abide by this order as there is no further mention of him in the press until Pius died in 1958.

Marriage

Until the nineteenth century, marriage had been a contract between a man and a man; a barter in which the merchandise was a young maiden. In parts of the United States it was common for a man to trade his daughter for a horse and believe he got the better part of the deal. The phenomenon of falling in love had very little to do with it. Marriage was a one-way street which purpose was to satisfy the lust of the man and grow his property, children. The woman was not much more than a limp rag in the bedroom.

In the mid-nineteenth century, when woman gained her recognition as a human being and was no longer seen as the mere property Moses had declared her to be, the falling in love syndrome started to become a factor in marriage. Yet, it still had to negotiate the hurdles society placed before it: creed, race, social status, and so forth. A Jewish girl who fell in love with a Christian man could not marry him. Even nationalities were involved. A Pole could not marry a Russian and so forth. Even if the engagement passed all the tests of society, it still had to pass the will of the family or the two could not marry. One had to choose between one's family and one's happiness. This was true of everyone, those at the top and those at the bottom.

Then there occurred an event that would ignite a social revolution which would forever change the definition of marriage. Marriage would no longer be the decision of others or for that matter the state. It would be the sole decision of two people who love each other. It would be an individual decision and not the decision of the majority. It would be the sacred duty of the state to recognize the individual commitment of two people who are in love - no matter who those two people were.

On December 10, 1936, Edward VIII, King of England, spoke to the people of England and the world, *"I abdicate my throne for the woman I love."*

Edward had been engaged to Wallis Simpson, an American divorcee. He had sought the approval of his family, the Church of England and the political establishment to no avail. Albino Luciani was twenty-four when Edward abdicated.

In Edward he saw great courage, the same kind courage his mother had shown when she had married his revolutionary father a quarter of a century earlier. In marrying a renegade, she too had given up most of her family.

Albino became an active part of this revolution. Having been of the *School of Einstein*, Luciani knew that the fundamental unit of all of God's creation was energy. He also knew the greatest manifestation of God's creation was mental energy, that is, the exchange of mental energy between human beings. He knew that every day people were exchanging mental energy with each other and there were some people who were mostly energy thieves, while others were mostly energy givers.

When one would tell new and interesting stories, one was an energy giver, whereas when one told repetitive and boring stories, one was an energy thief. When one said nice things to another, one was an energy giver and the gain of that energy could be profoundly felt by the receiver. Yet, on the other hand, when one yelled at or argued with another, one became an energy thief and the loss of that energy could be profoundly felt by the victim.

Luciani found there was only one circumstance in which the mental energy exchange could be perfectly even between two adults; when two people fell in love. Because it involved a perfect balance of the fundamental unit of God's creation - energy - he believed such a union of two minds could only be made by God and that Christ had been referring to this kind of union when He said, *"What God has joined together let no man put asunder."*

Unlike the Catholic Church might want to believe, Christ did not say, *"What preachers and other men have joined together let no man put asunder."*

Albino believed Christ sanctified the union between any two people who fell in love. Christ was clearly not referring to marriage as defined by other men or to marriage as made by other men or preachers. This gave Luciani clear conscience to perform and sanctify unions between people who previously been married by preachers. Something that he had been known to do quite often.

Prompted by the struggle in America in the early sixties to make marriage between blacks and whites legal, he made his position on the Holy Sacrament of Matrimony quite clear on a balmy afternoon in the summer of 1961. As presiding bishop, he was the keynote speaker at commencement services of the seminary in Vittorio Veneto. He summed up his lengthy address, in which he at times had gone into great detail as to how mental energy is exchanged between two people.

". . . Although the Bible's sole mention of the falling in love phenomenon describes it as occurring between two men: 1 Samuel 18, 'And it came to pass, when he had made an end of speaking unto Saul, that the soul of Jonathan was knit with the soul of David. Then Jonathan made a covenant with David, because he loved him as his own soul,' it applies to everyone. Therefore, we must hold most hallowed this solemn personification of God's creation - this perfect balance of mental energy that exists between two people when they fall in love - this perfect union of minds that can only be made by God.

"We must find the great courage within us to set aside the prejudice and hatred that has been implanted in us by our Christian forefathers and we must hold this kind of holy union in sanctified trust before Almighty God whenever it exists between any of God's children, whether it be between man and woman, or black and white, or Christian and Jew, or believer and unbeliever, or German and Russian, or royalty and commoner, or virgin and divorcee, or man and man, or woman and woman, or hermaphrodite and eunuch, or what have you.

"The rest of this thing we call 'love' is nothing more than the animal in us. To think differently - that it somehow pertains to physical parts of the body - is to say that the Holy Sacrament of Matrimony pertains equally to the apes in the wild as it does to human beings."

"When Christ said, 'Let no man put asunder what God has joined to together,' this is the kind of union He was speaking of. This union made by God. He was not speaking of the man-made unions we think of as marriage today." [20]

One must realize when John Paul said this, the movie *Guess Who's Coming to Dinner* had not yet been thought of and most of the other possibilities he spoke of were condemned by Christian preachers as being against the will of God.

It was clear to Luciani that the Church was in violation of Christ's will concerning marriage. In denying sanctity of marriage to any two people who were in love, the Church, blinded by prejudice, was *"putting asunder what God has joined together."*

Afterwards, a local reporter who had covered the event asked the bishop how he had been able to voice his opinion with such great conviction. How did he know that the phenomenon of falling in love could only be made by God? *"My father was the village troublemaker, particularly outspoken against the Church. The only Catholic part of him was the fact that he had been born and baptized a Catholic. When my mother, who was of an extremely devout family, fell in love with him and they approached the village priest he refused to marry them. In fact, my mother told me that they had to travel two hundred miles south to a small village on the Adriatic Sea where my father was unknown before they found a priest who was willing to marry them. She lost half of her family for having married a renegade, her brother and sisters never talked to her again. So this is why I know that this kind of holy union that is so powerful that it is able to bury centuries of prejudice and hatred of men can only be made by God."* [21]

An individual right

In the ensuing months, civil rights advocates in the United States, faced by population polls that were overwhelmingly against integration, continued their struggle state-by-state in their efforts to make interracial marriage legal. Luciani had led the same effort in Italy all the way to its Supreme Court. In that effort, he told the Italian magistrates,

"The state cannot tell its citizens who they can or cannot marry, less we cease to be a free society. Marriage is an individual right and not the decision of the majority. There is no issue more basic to freedom than is the right to marry whomever God deems

one fall in love with. There is no more fundamental issue that divides democracy from hypocrisy." [22]

Luciani objected to the American process in that it wrongly assumed that the right of marriage was not a basic human right and therefore was a state's prerogative. Even as late as 1967, when the United States Supreme Court finally rendered its decision in *Loving vs. Virginia*, there remained sixteen, mostly southern states, that did not permit marriages between blacks and whites.

Luciani criticized the American process in his address to the Christian Democratic Party Convention in 1963,

"In a free society, the most sacred duty of the federal government is to protect certain basic human rights for all of its citizens regardless of race or religious background and it should not be an option of any of its states or tributaries to abuse these basic rights. The right to fall in love with whomever God deems one fall in love with and the duty of the state to recognize such union is one of these basic human rights. These basic human rights are individual rights and therefore cannot be imposed upon by the state or for that matter by the majority. Democracy, which finds its strength in rule by the people, can only find its purpose, its sacred duty to society, in preserving the basic human rights of its loneliest individual." [23]

Priest and nun pedophilia

We have talked of Luciani's part in sex education in its role in eliminating family embarrassment which at his time was the leading cause of abortions. There was another reason for his work in sex education. His concern was the pedophile. He wanted to bring him out in the open where society could better keep an eye on him. He knew that if sex could be discussed openly within the family, abused children could better expose a transgressor and bring such abuse to an end.

He knew the percentage of priests who were pedophiles should be consistent with that of the general population. While studying for the priesthood he had done a paper[24] that addressed the question as to why the rate of pedophilia among priests seemed to be much higher than that of the general population and also why it appeared to be primarily homosexual while the preponderance of pedophilia was mostly heterosexual among the general population.

As already discussed, Luciani knew that most priests were either transsexual or homosexual because of the vow of celibacy. A straight teen had to make the great sacrifice in giving up sex for life because he had to give up the opportunity to marry and enjoy a life of sex free of sin whereas the gay teen didn't have to give up a thing, as for him marriage was not an option. He was condemned to live a life of celibacy anyway.

Most transsexuals - women born into men's bodies - are perceived by the general public to be homosexuals and this is because few of them op at surgery and instead live a homosexual lifestyle. Being women, they have the natural instinct to fall in love with men and since they have men's bodies they are perceived to be homosexuals whereas, in fact, in their case, the problem is one of sexual identity and not one of sexual orientation.

Actually, most transsexual men believe themselves to be effeminate homosexuals. They live their lives unaware that their problem is one of identity and not one of orientation. Transsexual men, having women's minds, are naturally effeminate; this causes another misconception on the part of the public. It perceives most homosexual men to be effeminate.

In truth, although it is possible for a man to acquire effeminate traits in this life, very few homosexual men ever do. This is because a male homosexual accepts himself as a man. He does not think of himself as a woman, otherwise he would be a transsexual. The latest statistics tell us that almost ninety percent of the homosexual population remains underground. As more and more gays come out of the 'closet' the public perception will change to one of realizing that relatively few male homosexuals are effeminate.

Regardless, Luciani recognized an inconsistency, priests versus nuns and monks, *"I am particularly puzzled as to why there is practically zero incidence of pedophilia among nuns and monks who outnumber priests four to one and who in their respective roles as teachers and counselors of youth programs have immensely greater access to children than do priests. Certainly,"* he wrote, *"all three professions would attract the same mix of people as they offer one a life of celibacy, one dedicated to God. I wondered what could be the difference between a priest and a monk or a nun that could result in such a high rate of pedophilia among priests."*

He researched those who had gone before him who had looked into the problem in the past. He found one researcher who believed that a part of the problem was caused by the vestments themselves, because they were attractive to the transgender population. Certainly, a monk's brown robe would not suffice, as everyone else would be wearing the same dress. And certainly the reverse transgender population - men born into women's bodies - would have no interest in the convent, as at the time it required that they spend the rest of their lives in a black dress. This segment of the transgender population wanted to wear pants, not dresses. The researcher had concluded that this could result in a significant transgender population among priests and zero transgender population among monks and nuns.

This conclusion has some truth today. Although the transgender population is relatively small, it is relatively large when compared to the priest population. For example, there are over five million transsexuals in the United States and there are less than fifty thousand priests. It is reasonable to believe that attracted by the vestments themselves there could be a significant percentage of transsexuals among priests. Particularly, that like a homosexual, a transsexual in the vow of celibacy doesn't have to give up anything. Being unable to marry in the eyes of the Church, he or she would otherwise be condemned to a life of sin.

There is a marked difference in pedophilia as it pertains to homosexuals versus transsexuals. When a homosexual priest has sex with a male child it is a homosexual act. When a transsexual priest has sex with a male child it is a heterosexual act as the transgressor is in fact a woman trapped in a man's body.

Luciani reasoned that it made some sense that there existed an abnormally high number of transgender people among priests, women in men's bodies, and he concluded that this could result in a high rate of 'homosexual' pedophilia activity when one considers that the only significant contact priests had with children at the time was with altar boys. Because the general public assumes that a transsexual is a homosexual, this would be heterosexual activity that would be perceived to be homosexual activity as it would involve women who were trapped in men's bodies.

As just mentioned, most people view transsexuals as being homosexuals. Moreover, most transsexuals mistakenly believe they are homosexuals, particularly in their early years. In actuality, the great majority of transsexuals are heterosexual, the reason they never marry, as believing they are of the opposite sex they can only fall in love with one of their own physical sex. The few transsexuals that do marry are homosexuals. A woman born into a man's body, for example, who is a homosexual, can fall in love with another women, and since 'she' has a man's biological body 'she' can father children. *"This could explain a part of the problem, why priest pedophilia appeared to be primarily homosexual in nature. But not good enough,"* Luciani surmised, *"There must be something else?"*

The only known cure for pedophilia at the time, which is still true today, was to keep the pedophile away from situations in which he or she is alone with a child. He knew that a priest was largely limited to altar boys, whereas a nun had unlimited access to all children. This is still true today as the altar girls of today are usually under the supervision of nuns. He also knew that an innocent act on the part of the child usually excited or aroused the pedophilia instinct in the adult and that this combined with access to the child resulted in the transgression.

Luciani traced much of the problem to the confessional box. The confessional box explained why there was such a high incidence of pedophilia among priest and yet a low incidence of pedophilia among nuns and monks.

He knew that eight, nine, ten and eleven year old children often talked about their 'sins of the flesh' to the priest in the confessional. *"After all, masturbation is a sin,"* he wrote, *"Actually, by doctrine, a mortal sin in the eyes of the Church. Children, lacking a vocabulary to otherwise explain their actions, often go to great detail in discussing what they do with their sexual bodies. This can be a very dangerous thing. I find that the record shows that pedophile priests who otherwise would remain dormant are often aroused by these children's stories. In addition, I find that the pedophile priest often uses the confessional box to sort out those children or teens that were more likely to be receptive to his advances, which makes easier prey of them."*

He concluded that the confessional box was a great weapon in the arsenal of the pedophile priest and he intended to take it from him, to eliminate to a great extent the cause of the problem. In these things, Luciani recognized the basic difference between sexual orientation and pedophilia. He knew that sexual orientation is something that one is born with - it determines who one can fall in love with. Pedophilia, on the other hand, is a sexual perversion - a sexual fetish. It is acquired as an adult.

Luciani made this clear, *"No one is born with the instinct to have sex with children. Falling in love has nothing to do with having sex with an eight year old child. No adult has a romantic affair with an eight year old child. Who one is able to fall in love with - sexual orientation - has nothing to do with pedophilia."*

This told him that purging the clergy of homosexual and transsexual priests would not solve the problem. It would only shift the problem to the heterosexual arena. The overwhelming number of pedophilia cases in society involves incest; a parent has the greatest private access to children. Eliminating homosexuals would leave heterosexual pedophile priests with private access to children in the confessional box.

He resolved that no child short of puberty should be subjected to the confessional, that they be permitted access to the Eucharist until the age of fourteen without the requirement of confession. A few years later, as a common priest, he appealed his cause directly to Rome, but all it brought him was the chastisement of his bishop for having broken protocol. Once again, his voice had gone unheard. He would store this one up with all the others for that day on which he would someday be infallible.

Nevertheless, in his own jurisdiction, whether it was a small parish, a diocese or an archdiocese, he preached the most prolific testimony of his ministry, *"We have made of sex the greatest of sins, whereas in itself, it is human nature and not a sin at all."*

Tried criminal cases today confirm Luciani's conclusion that the confessional booth is at the core of the problem. The combination of the Church's position that all sex outside of marriage including masturbation is a mortal sin, together with the requirement of confession, has played a role in most of them.

Regardless of what individual priests might tell their parishioners, any kind of sex outside of marriage is a mortal sin according to both the Bible and Church doctrine.

The confessional box remains a cesspool of sexual perversion even today as many Catholic children believe unless they confess their sexual transgressions to a priest they will never get into heaven. The typical priest is also bombarded by adults week-in and week-out with endless tales of a range of sexual transgressions and sick and perverted practices, from sadistic and masochistic rituals and rape and incest and even sexual murders.

My friend Jack once told me that although the vow of chastity is a challenge to all those who take it, a priest, a monk and a nun, it is often an insurmountable one to a priest, particularly a young priest. *"To begin with, they brainwash you with the word purity throughout the seminary years and then they ordain you and put you inside the pressure cooker, the confessional box, and force you to listen to pornographic tales that stagger the imagination for hours every Saturday afternoon. It is the equivalent of starving a man to the brink of death and then waving a mouthwatering porterhouse steak in front of him."*

So one can easily understand why so many priests, being constantly exposed to an ongoing parade of livid pornography of people's everyday perverted sexual lives, do themselves become perverted; some, confining themselves to self-manipulation in the privacy of their booths, while others become aggressive perverted monsters of prey.

One might wonder why the Conference of Bishops, when it met in 2003 concerning the issue of priest pedophilia, never mentioned the confessional as being a factor in priest pedophilia. Also, one would wonder why, with actual tried criminal cases pointing overwhelmingly to the confessional box, no judge has yet issued an injunction against the Church's practice of confession. Where most might think that this is a matter of separation of church and state, or respect for the papacy, it is most likely to be a matter of lack of courage. Society has the sacred responsibility to protect all of its children until they reach adulthood even when the wrongdoer is the family. We see this in everyday life; parents influenced by religious beliefs are tried and incarcerated for endangering the physical or mental health of their children.

In the face of mounting priest pedophilia charges, early in 2004, Cardinal Winning, Primate of Scotland, made a change that prevented children from receiving their First Communion before they made their Confirmation. This deferred their first confession to the age of fourteen. In order not to draw the wrath of Rome, the cardinal used the reasoning that he was simply returning to what had been the practice centuries ago. Since he made the change, no priest in Scotland has been charged with pedophilia, that is, molestation of a pre-adolescent child. Hebophilia, molestation of a teenager, is not pedophilia.

It should be pointed out that hebophilia, unlike pedophilia, is usually a matter of sexual orientation particularly if the victim is in the late teens. Unlike pedophilia, hebophilia is a matter of definition as determined by the country it takes place in. The age of consent without parent's permission is sixteen in most countries. In the United States, it has remained at eighteen, as it was fifty years ago, despite the fact that today's sixteen year old knows immensely more about sex than did his counterpart of twenty-one fifty years ago. No sixteen year-old has sex with an adult without knowing what he or she is doing.

If a teenager gets his or her parent's consent, the bar drops dramatically. For example, in the state of Massachusetts, today, the age of consent is twelve with a parent's permission.

In the spring of 1978, just a few months before his election, Luciani was reprimanded by the Vatican for having criticized an American bishop who had paid off the alleged victim of a pedophile priest. He told a group of bishops that had gathered in Venice, *"It would be better that we try our accused fellow servants in a court of law so that they can be cleared of any wrong doing and if found guilty they should pay their debt to society. It is not Mother Church's business to pay their debt in cash, particularly to pay it with money that was intended for the poor. Besides, if we take no action to get at the truth, we may very well be endangering countless children in the future."*[25]

The seal of the confessional box

Even more abhorrent to him as a young priest was that on occasion a child of seven or eight would confess to him of an incestuous affair with a parent. Also, in one case, on repeated occasions, a certain priest had confessed to him that he was a serial pedophile. He knew that all over the world this must have been a tormenting thing for priests who, bound by the seal of the confessional box, were unable to bring an end to this kind of atrocity. He voiced the problem to Rome and called that some modifications be made to the strict guidelines of the confessional imposed by the Church. He argued that it was a mortal sin for a priest to conceal abuse of a young child and to permit such an abomination to go on. This time he received a formal response from the Vatican, *"God will take care of those things. The confessional box must remain sealed."* But Luciani knew in his heart that God would not take care of those things; that it was up to him to take care of "those things." And he vowed in his memoirs, *"Someday I will take care of 'those things', once and for all!"*

The Sideshow of the Roman Catholic Church

In 1973, Albino Luciani was named a cardinal and two years later in July of 1975 he launched his most vigorous attack in the media concerning any issue of his ministry. He severely criticized a South American bishop for capitalizing on a comatose twelve year old girl who was alleged to have the stigmata. The girl had been injured when she was nine in a farming accident and had never regained her faculties. The local bishop and the parents of the girl put her on public display through a one-way window which had been set into the wall of her bedroom. In a circus-like atmosphere they paraded thousands of pilgrims past her room for a substantial fee; enough so that the bishop was able to build a new mansion and the parents became millionaires.

His release, *"I can understand why Rome might turn her head the other way when greedy people capitalize on defenseless children in this way. After all, she is at the source of this kind of satanic ritual. What bothers me most is that men and women of good conscience of the state stand aside and do nothing about it."*[26]

Stigmata and the Shroud

Luciani, like most educated people today, knew stigmata had been a fraudulent money-making scheme in the Church ever since the thirteenth century when St. Francis had been the first to display wounds on the palms of the hands. Strangely, there is no record of a claim of stigmata before St Francis; he is rightfully credited with having come up with this perverted idea. All popes since his time have capitalized on this *Sideshow of the Roman Catholic Church.*

Luciani knew crucifixions involved the driving of spikes into planks of wood placed over the outstretched arms and then through the upper-arms and forearms just beneath the bone line of the victim and into the cross. It wouldn't take a Roman historian to tell one that; and it would not take a PhD in Structural Engineering of the Anatomy, to tell one that nails, particularly razor-sharp-edged first century nails, if driven through the palms of the hands, would never support the weight of a hanging human body.

This had been well publicized a short time before the South American incident, when a photograph of the Shroud of Turin disclosed that the blood stains on the shroud evidenced the blood was coming from the wrists and not from the palms; Christ was crucified through the wrists and not through the palms. Of course, one knows that Christ could not have been crucified through the wrists for He would have bled to death in a few minutes as the main arteries located there would have ruptured under the weight of His body upon hanging.

The pain in crucifixion was concentrated in the feet. Spikes were driven crisscross just under the ankles into the cross. The hanging would severely impair the victim's ability to breathe and in order to keep breathing the victim would have to put pressure on his heels which caused unbearable pain. In the end the victim would succumb to asphyxiation.

The official position of the Church concerning the shroud is it is a fake. When it first appeared in a small village in France in 1353AD, Pope Innocent VI ordered an investigation. After a three year look into the matter, Bishop Pierre d'Arcis of Paris reported, *"After diligent inquiry and examination, the truth attested by the artist who created it, to wit, that it is the work of human skill and not miraculously bestowed. Bishop Henri de Peituers falsely and deceitfully created the hoax for personal gain . . . In his deception, Bishop de Peituers had gone so far as to secure a cloth woven in the Holy Land."*[27]

The pope ordered it removed from display. After Innocent died, however, it resurfaced across Europe. His successor Urban V, realizing that it was good for business, allowed the practice to go on. Yet, no pope since has overruled Innocent's decree.

Today, the official position of the Church concerning the placement of the nails is based on the photographer's discovery; Christ was nailed through the wrists. Yet, the Church continues to recognize stigmata only if it appears in the palms.

Author's collection

Shroud 1353AD   Relief 1119AD   Copy 2003AD

Above is a sculptured relief over the main entrance of a French church. Lower-left is the shroud. Lower center is an enlargement of Christ's face in the relief. Lower-right is an image retrieved in 2003 by scaffold workers by applying powdered resin to the sculpture and placing a cloth over it. The church was consecrated in 1119AD.

The image itself differs slightly because of aging and a large crack that occurred since the fourteenth century and the way the cloth was placed to secure the image. Yet, the hair pattern on either side of the image is an exact match, strand for strand.

Although they differ as to the shroud's authenticity, experts on both sides of the aisle agree the face in the shroud is definitely French. If the shroud is authentic, Jesus Christ was French.

Luciani once told me, "Scrupulous men continue to prey on the great weakness of man - his mortality. Particularly vulnerable are those that would grasp at any straw that could possibly be connected to their hope for immortality. Take a peek at Jewish custom, as to how the Jews buried their dead at the time. While the body was customarily wrapped in a white linen shroud as the Bible explicitly states in the *Gospel of John* 20, *'Then they took the body of Jesus, and wound it in linen clothes, as the manner of the Jews is to bury,'* the face was first wrapped in a linen napkin as told in the same Gospel, *'Then cometh Simon Peter into the sepulcher, and seeing the linen clothes lie, And the napkin that was wrapped about his face, not lying with the linen clothes, but wrapped together in one place by itself.'* Where our fourteenth century hoaxers went wrong is that they failed to check out the Bible, which tells us explicitly that if the image of the face of Jesus survived today, it would be on the napkin, and not on the shroud."

The Miracle of Stigmata

Luciani never believed in stigmata despite the fact more than sixty people had achieved sainthood principally because they had claimed the affliction. He knew that Christ had in fact been crucified through the arms as both the history books and common sense tell us. He also knew that when artists, several hundred years after Christ's death, painted renditions of the crucifixion, they had placed the nails in the palms because it was more artistic. It was this that resulted in the misconception ever since.

Furthermore, he knew the Church knew this. It was obvious to Luciani that if stigmata were truly a miracle - as the Church claimed it was - the wounds would appear in the biceps and forearms. Otherwise, one would have to conclude that God the Father, who bestowed this affliction on certain people, had a short memory of what had actually happened in the crucifixion and has to depend on more recent artist renditions.

Every day, thousands of the billions of people in the world develop cancerous and bleeding sores and ulcers in various parts of the body, sometimes in the toes, sometimes in the genitalia, sometimes in the buttocks, sometimes in the knees, sometimes in the neck, sometimes on the face, and so forth. Also, stigmata can be a manifestation of hysteria and other mental disorders.

Nevertheless, if stigmata were in fact an act of God, then the Church should have been granting sainthood to those afflicted in the upperarms and forearms. Of course, the Church has itself found the vast majority of stigmata to have been self-inflicted; in the case of the comatose girl in South America, it was torture.

The situation was handled as quietly as possible behind closed doors in the Vatican, as Luciani's harsh criticism rippled across Latin America. Paul VI ordered the local bishop to stop the 'circus' and for all intents and purposes that was the end of it. What's more, Paul took measures to bring an end to several similar atrocities that were taking place elsewhere.

Normally this would have been a milestone event in the Church's history as it would have marked the beginning of the end of mysticism in the Roman Catholic Church. However, when John Paul II rose to the papacy he gave a green light to these things provided they were under the supervision of the Vatican which meant that the Vatican would get the lion's share of the proceeds.

One of the most famous of these involves Audrey Santos, a comatose teenage girl in Worcester Massachusetts who has been displayed before thousands of people since she had a pool accident more than a decade ago. What's more, he canonized a number of saints solely on the basis of the 'miracle' of stigmata.

The mysticism of sainthood

In general, Luciani had a great problem with the way the Church determined candidates for sainthood. He felt there should be a single yardstick for sainthood: the person must have made a substantial contribution toward making this a better place to bring the world's future children into. He had studied the lives of all the saints and he knew that none of them were in this category; as a matter a fact most of them had during their lifetimes made this a worse place to bring the world's future children into.

In fact, he found that many of the 'saints' had never even lived; that is they were fictitious characters someone had written books about. This included all except a half dozen of the three dozen popes listed by the Church in its claim of apostolic succession between Peter and Sylvester, the first bishop of Rome appointed by Constantine three hundred years after Christ's time.

The Church's requirements were mainly that the candidate for sainthood spent his or her lifetime on his or her knees with the additional requirement of magic; the candidate had to have been involved in two miracles. To Luciani, a miracle was a trick of magic. According to the rules of the Church a person who made an enormous contribution to mankind could never be considered for sainthood unless he or she had performed magic along the way.

He knew that everyday thousands of cases of cancer for no known reason would go into remission. If one of these fortunate souls had happened to have said a prayer to a deceased man of the Church that man became a candidate for sainthood. The great lion's share of 'miracles' that supported the hundreds of saints canonized by John Paul II, including his own canonization, rely on cases of cancer going into remission.

A grain of sand on the beach

Luciani recognized that the ideology of medical miracles contradicted the basic premise of religion. Religion tells us that this life is as a grain of sand on the beach of eternity. It tells us the only purpose of this life is to get us to the afterlife. Medical and other miracles are sought by those who don't really believe in an afterlife.

As a bishop he told me, "It makes no sense to pray for the recovery of an eight year old boy who is suffering from cancer when if he dies one knows he goes straight to heaven. Christ said, *'Blessed are the little children for theirs is the Kingdom of Heaven.'* Christ said this many times as to leave no doubt about it. If the child dies he has an absolute chance at heaven; if he grows to adulthood, he reduces his chances immeasurably. *"Many are called, few are chosen."* If one believes in the gospels, if one believes in religion, there is no point to miracles in this life?"

I gave him a dumbfounded glance.

He answered my unasked question, "The point is clear. Those who pray for medical miracles in this life don't really believe in an afterlife. People who believe in miracles are those who are too weak to accept the will of the *God of Nature* – the natural order of God's creation - the God of the atheist. The God we know, as a matter-of-absolute-fact, gives us life."

Lourdes

Anyone who is aware of the least of his life knows Luciani did not believe in the many apparitions of ghosts who claimed to be God in ancient times; Moses and the other prophets. He believed God comes to each of us in our conscience.

This extended to modern apparitions. In 1964, on the guise of pilgrimage, he visited Lourdes. In trying to resolve the issue for himself, he gained access to original transcripts of the events in the diocese office. He determined there were two decisive issues in the events that had made the case for the Church.

The first was the fact that a spring had sprung forth in the grotto where the 'lady' appeared to Bernadette - a miracle. Yet, he found in the original transcript the first apparition occurred when *"Bernadette went into the grotto to fetch water for the day."* That the spring was always there was confirmed by testimony of townsfolk. Nevertheless, the Church claimed a new tributary of the spring had appeared.

The second issue, the more critical one, was that the 'lady' told Bernadette that she was the *Immaculate Conception*. At the time, the *Doctrine of the Immaculate Conception* - Mary had been born without Original Sin - was not known among the general congregation. The doctrine had been drawn up in 1854 at the direction of Pius IX, just four years before the alleged visions.

Like most progressive doctrines before it, this doctrine was enacted to bring the Church into sync with social changes. For centuries, Christian nations recognized women as mere property as decreed by Moses. In 1853, following Europe's lead, laws were passed in the United States elevating women from being mere property to citizens bringing pressure on the Vatican to elevate the concept of women within the Church. In 1854, Pius IX came up with the idea of the *Immaculate Conception*. Yet, he was faced with the fact that for eighteen centuries Mary was assumed to have born with Original Sin, and suddenly the Pope found out she had been born free of the sin. He decided to keep the doctrine secret as he searched for a way to convince the public of its authenticity.

It was that Bernadette spoke of this doctrine that made her a saint. There was no way she could have learned of this doctrine. Luciani left Lourdes with the conviction the spring had always been there. Yet, he was still puzzled as to how she could have learned of the *Immaculate Conception*.

In 1971, he returned to Lourdes. This time he spent a week scouring over records and talking to a range of witnesses.

The first apparition occurred in January 1858 and the following fifteen apparitions were witnessed by crowds. The only words witnesses heard her utter were "Aquerdo" – "That's she." She would tell the crowd she had seen a girl of her own age who was silent as to who she was. Bernadette was fourteen.

Bernadette was interviewed by three dozen clergy, mostly bishops. In early March, just the day before the sixteenth apparition, she was interviewed by two bishops from the Vatican.

The sixteenth apparition was witnessed by ten thousand people. No one saw anything except a young girl talking to the trees. After the vision, Bernadette told the crowd, the young girl had told her, *"I am the Immaculate Conception."*

The next day, a row of several dozen crutches appeared lined up against a wall; suggesting crippled visitors had walked away. Luciani interviewed a number of the townsfolk who told him the crutches were placed there by greedy villagers who saw a chance to capitalize on the situation. It worked. Within months, the once poverty stricken town was booming

There was no greater feminist activist in the history of the world than Albino Luciani, himself. One could ask why he would destroy the myth of the *Immaculate Conception* when it would contribute toward his objective - to bring about the equality of women in the Church.

He would pull Mother Church's head up out of the sand. He would not be a part to a deception designed to fool people into thinking women are equal. He would remove the obstacles the Church and society had placed her path. Woman, herself, would prove she was equal.

Fatima

In 1977, Luciani looked into the other apparition of the times - Fatima. Seventy thousand people and hundreds of reporters witnessed the event of October 13, 1917. The three children who claimed to have seen the 'lady' were Lucia dos Santos age 10 and her cousins Francisco Marto age 9 and Jacinta Marto age 7.

The next day, the Lisbon *O Dia* reported, *"The silver sun, enveloped in a gauzy grey light, was seen to turn in a circle of broken clouds . . ."* Its allied newspapers *Ordem* and *O'Seculo* reported a similar sighting. With thousands of cameras at hand, no one thought of taking a picture of *the miracle of the sun.* Three hundred European newspapers made no mention of it. A week later, tabloids began to exaggerate and capitalize on the 'sun' story.

A year after the apparition, newspapers reported the boy Francisco inconsistent in his story. A week later he fell ill and died a few months afterwards of a severe urinary infection.

Lucia had been the spokesman in the visions; the attention and money centered on her. Although there is no record in the press, it is possible that Jacinta demanded a part of the action as shortly after her brother's death she fell ill and suffered for a year with stomach pains before dying of a bladder infection. One could assume the children died of the Flu which killed one-half-of-one-percent of the population two years earlier. The Vatican's beatification process claims the children suffered excruciating abdominal pains. The Flu was a quick deadly killer ending in respiratory failure; not symptomatic of great pain. Cancer would be the first thing a doctor would suspect. Yet, painful infection of digestive and purification organs is also symptomatic of slow arsenic poisoning and other poisonings.

The Vatican had capitalized on Fatima. It would destroy its credibility should the younger children deny the apparition; the infallibility of the papacy was at stake. When it comes to murder, the Vatican may not stop at popes.

It could be coincidence that two of the three children who witnessed the apparition died at about the same time shortly after the event. The mathematical possibility would be about one in ten billion. It could be that God, knowing these children were a danger to Rome, took them back.

Yet, it is possible Lucia, or her parents, killed the children. Lucia had already gained world fame and her family had been raised from poverty to great wealth overnight which they stood to lose if the younger children survived.

On July 9, 1977, Albino Luciani interviewed the aging Lucia at Coimbra. Scores of reporters swarmed the cardinal as he left the convent. All they got was a look of anguish and despair.

Lucia dos Santos

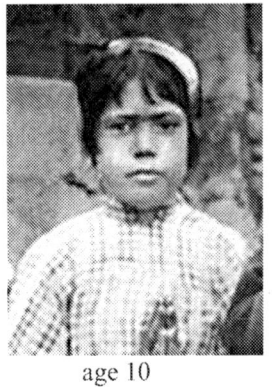 

age 10   age 40

Truth or Deception?

It is certainly an occurrence of extraordinary proportions that of the more than two thousand apparitions recognized by the Church, the only one witnessed by more than one person, two of those persons died within a very short time at a very young age.

Archbishop by popular demand

In 1958, Pope Pius died and Luciani, freed of restrictions this Pope had placed on him, *came out on all fours*. Almost immediately, John XXIII sharpened his claws by making him into a bishop, a prince of the Church.

For the next few years the Bishop of Vittorio Veneto served as a leading advisor to John XXIII and Paul VI, the two popes of the twentieth century who brought change to the Church. John once said of him, *"The prince of the Veneto country is Lacordaire and Rosmini in one. Today I follow in his footprints in order that someday he might follow in mine!"* The Pope was referring to two of the greatest theologians in Church history.

That John alluded to the possibility Luciani might succeed him confused his audience as it was generally assumed that John had intended Giovanni Montini to follow him as Pope. After all, one of the first things he had done upon becoming Pope was to name Montini a cardinal. Since John himself had some years left and Montini seemed good for an additional twenty or so the papacy appeared to be locked up for a quarter of a century.

In that John had named a second successor could mean only one thing. He knew that Montini, who silently shared many of Luciani's liberal views, would be in danger in the Vatican; that Montini, elected Paul VI in 1963, might not live out his natural years.

In 1969, the Archbishop of Venice died. This post is the most coveted field position in Italy because it is the only one that holds the title of Patriarch; one of only two such positions in the Catholic world, the other is the Patriarch of Lisbon. Historically, for this reason, a ranking cardinal had normally been moved from the Vatican to fill a vacancy.

Realizing a Vatican cardinal would be chosen, thousands from the Veneto country marched through the streets of Venice demanding the bishop of the remote mountain diocese of Vittorio Veneto be named to the post.

When Luciani was elevated to the position, the press billed it as the first democratic election of an archbishop in the history of the Church. But as we shall see in what is yet to come, Paul VI had already chosen Albino Luciani for the job. As a matter-of-fact he had already chosen him to one day fill his own job.

His defenseless friends

When one considers the oppressed, Luciani had one more on his agenda; not one, should he rise to the papacy, he would use his dictatorial powers to accomplish. Yet, nevertheless, it was there. He would employ the process of education and awareness and let conscience do the rest of the job.

He would tell his parishioners the truth, one they would rather not know about. How long a calf lived had become a matter of economics. Nations had enacted laws that governed the selling of meat; a calf killed before its four-month birthday could be sold at premium calf prices; one killed after that would go for lower beef prices. It would be a short reprieve for the survivor, as he would meet his fate a few months down the road. It would not make economic sense to continue to feed a calf beyond that time as the benefit of incremental girth at lower prices for less tender meat would not justify the cost of incremental grain.

Much to the chagrin of cattle ranchers and dairy farmers, he would counsel children to this end. He would tell them, *"Paying another man to kill another man doesn't absolve one of guilt."*[28] As a bishop and as a cardinal, he was often pictured with children and young farm animals.

Luciani with his creatures, great and small

White Light Dark Night

Luciani believed, when men became aware of this injustice, their compassion would take hold and they would return to be the vegetarians God had willed them to be. He believed in time, perhaps not in his lifetime, the children of the animal world would follow their counterparts, the children of the people world, into the world of equality; many who, too, had struggled for their rights.

**Life, Liberty and the Pursuit of Happiness
. . . all creatures, great and small**

Just a regular guy

It might be of interest to the reader to know that during his tenure as Archbishop of Venice Luciani lived as a common man rather than as the crown prince of the Church that he was. Often, he would go incognito through the towns and villages that made up his diocese dressed in clothes not much more than those befitting a bum. He would learn first hand the needs of his congregation. This together with that he came across as a regular guy rather than a man of great stature occasionally got him into trouble.

In one case he was stopped by police in an impoverished section of a village and asked for his identification. When he failed to produce his papers and claimed to be the Patriarch of Venice they didn't believe him and placed him under arrest.

On another occasion he entered into a conversation with a group of students in a pub and invited them to come and see him at home on the morrow. They asked him where he lived. "On the Piazzetta dei Leoncini," he replied. They asked, "You mean the Patriarch's Palace?" "Yes," he replied. "And for whom should we ask?" they queried. "Just ask for Piccolo," he told them.

The next day when they arrived at the palace and asked for Piccolo they were taken to an elevator flanked by Swiss guards which took them to the fifth floor where they were led to an office. When they entered the room they were astonished to find Piccolo sitting by the window chatting with Pope Paul who happened to be visiting him that day.

Luciani despised wearing vestments as he thought there to be something hypocritical about them. One day, in the park that fronted the Patriarch's Palace in Venice, dressed in shorts and sandals, he was asked by a severely handicapped boy to pose with him for a photograph.

When he motioned to kneel down beside the boy, the boy stopped him and looking up at him asked, *"Could you put on your beautiful clothes?"* Luciani got up and disappeared into the palace. He returned wearing his cardinal's robe, miter and golden staff. *"There are times,"* he told the boy's father, *"we must shed our humility and put on our hypocrisy, and this is one of them."*[29]

White Light Dark Night

[1] pg 84 Albino Luciani, Italian Parliament 16 Jan 59
[2] pg 84 Albino Luciani, Italian Parliament 16 Jan 59
[3] pg 85 Diocese Belluno/ Diocese Vittorio Veneto
[4] pg 86 Diocese Belluno/ Diocese Vittorio Veneto
[5] pg 86 L'Osservatore Romano 29 Mar 78 Nostro
[6] pg 86-89 *Strategy of a Strange War* March 1941, Apostolic Library. Not to be confused with his doctoral thesis, Origin of the Human Soul. February 1947. All *italicized* text is reprinted from his 1941 thesis *Strategy of a Strange War*.
[7] pg 91 Veneto Nostro 16 Jul 66
[8] pg 91 The court reversed its decision in November 2003
[9] pg 91 Statistical Abstracts of the United States 1960-1979
[10] pg 92 Veneto Nostro May 69
[11] pg 94 Il Gazzattino Venezia 14 Jul 76
[12] pg 95 Il Gazzattino Venezia 23 Jul 76
[13] pg 96 Nostro Provoilego "Our Privilege" 12 Apr 70
[14] pg 96 visit websites of right wing enemies of John Paul I who remember what he was about. E.g. The Scandals and Heresies of John Paul I by Brother Michael Dimond O. S. B.
[15] pg 98 IL Gazzattino Venezia 21 Jul 78 and in world newspapers
[16] pg 100 Messaggero Mestre 2 Aug 78
[17] pg 100 Princeton Packet 2 Apr 49
[18] pg 101 Secolo Genoa 21 Jul 49
[19] pg 103 Veneto Nostro 17 Jul 51
[20] pg 106 Messaggero Veneto 24 Jun 61
[21] pg 107 Veneto Nostro 28 Jun 61
[22] pg 108 Bishop Albino Luciani Italian Supreme Court 22 Aug 59
[23] pg 108 Christian Democratic Party minutes 22 Aug 63
[24] pg 108 Diocese of Vittorio Veneto previously Diocese of Belluno
[25] pg 114 Messaggero Mestre 17 Mar 78
[26] pg 115 Messaggero Mestre 12 Jul 75
[27] pg 117 Apostolic Library, the Vatican
[28] pg 126 Messaggero Veneto 21 Jun 70
[29] pg 128 Jack Champney witnessed this in Venice on March 9, 1975 and told the author of it shortly afterwards. Sister Vincenza retold the story a decade later to several magazines. These conservative publications edited out Luciani's comment concerning the hypocrisy of vestments.

Chapter 8

Two Very Different Kinds of Popes

In these pages I have put before you and what is yet to come, I have and I will often refer to John Paul II, as he was the successor of my patron John Paul. In the event you have not already picked it up, these were two very different kinds of men, one an ultraliberal and the other an ultraconservative. How far apart were these two in their basic ideologies and why did they think so very differently?

First, one must consider their upbringing. Karol Wojtyla was born into money so, unlike Albino Luciani, he never knew what it was to wonder where his next meal was coming from. Also, unlike Luciani, he was born to two devout parents who were steadfast believers in whatever the old men in Rome happened to tell them. From the age of six through the time he entered college, Karol was educated in expensive prep schools, something that even today is reserved for the very, very rich.

Karol Wojtyla and his mother

Eight year old Karol Wojtyla is shown with his mother wearing the military uniform of the Wadowice Military Academy, a prep school that brainwashed its students in *Fascism*, that ideology which preached the superiority of the white Aryan male and the subordination of women, Jews, blacks, homosexuals and certain ethnic groups. While he was growing up, in the interests of preserving the purity of the Aryan race, blacks were not allowed in Poland and in other predominately Catholic countries in Europe.

John Paul II has been unfairly criticized in the media because as a youth he refused to join the underground resistance during the war and instead went to work for the Nazis.

In Krakow, sixty thousand courageous youths escaped to the sewers, Karol Wojtyla was not among them. The reason he was not among them was that they were overwhelmingly Jews and Jew supporters. They knew what *Nazism* stood for, after all *Mein Kampf* had spelt out their pending doom quite clearly.

Yet, there was a more telling reason Karol went to work for the Nazis. He, like most Polish youths, shared the same ideologies as did Hitler and Pius XII - the superiority of the white Aryan male, the subordination of women, the segregation of blacks and the persecution of homosexuals, Jews, Muslims and others.

One might question why Poland fell in only a few days? After all, it had one of the most formidable and best prepared armies in Europe. In addition, it had by treaty the promise of both Britain and France as allies should Germany invade Poland.

The Poles and the Nazis shared the same ideals. Their common enemy was Communist Russia, not Fascism. In Poland, an army of twenty-five million fascists eagerly went to work building ammunition and supplies for the German army while only a small handful joined the resistance. Karol Wojtyla like everyone else was caught up in the *Opium of the Masses, Christianity.*

Onward Christian Soldiers

Motion pictures and books that portray German troops invading and shooting up churches in Poland could not be further from the truth. This may have been somewhat true of the Russians, but not of the Germans. This is modern day propaganda designed to separate modern Christianity from what it was then - *Fascism*.

When the war began, by a whisker over Italy, Germany was the most Christian country in the world split between Catholics and

Protestants. Its ally, the Italian army, was entirely Catholic. In those days everyone was devout; everyone went to church.

Hitler's Guard attends the *Holy Sacrifice of the Mass*

Unlike the allied armies which spanned many religions, the axis military was entirely Christian. There was not a Jew, a black or a Muslim in the German or Italian armies. The Axis powers had the largest chaplain core of any army in history The *Holy Sacrifice of the Mass* and other services were available to members every day. The hymn, *Onward Christian Soldiers*, could be heard in both German and Italian for miles surrounding battlefields.

Hitler leaving mass

White Light Dark Night

A Nazi Chaplain in combat

Battleground graves of fallen Nazi soldiers

Driven by a common cause, Karol quickly progressed from a laborer in a quarry to a supervisor position in a chemical supply depot of the German army. Karol also joined a small theater group. The love of his life was the theater. Above all he wanted to be an actor. The record shows that he was quite a successful actor, usually commanding the leading role. One could ask, if Karol wanted to be an actor and had no interest in being a priest, why he entered a seminary?

Karol Wojtyla in a romantic satire of Hitler's youth (1942)

As the war progressed, growing expectation of an allied victory was forcing Polish youth into the German Army. This was particularly true of young men like Karol who were rising in the German ranks outside the army. In both the German and Italian armies, enrollment in a seminary was cause for deferment even though fathers of young children were not excused. Toward war's end, even young children were not excused. He entered a seminary and soon realized he could use his acting abilities in the Church.

This is why he has been the best imaged pope in history. What the public has seen is the epitome of a great actor on the world stage backed by a multi-million dollar public relations budget. It is unlikely his successors will match his performance unless one comes out of Hollywood. For almost a decade, the two most powerful men in the world were actors, John Paul II in Rome and Ronald Reagan in Washington.

Conversely, Ratzinger fought valiantly for his fatherland, deserting late in the war when to continue was obvious suicide.

Nevertheless, the difference between these kinds of men was the method by which they would attain their mutual objectives - *Hitler and Mussolini* through annihilation versus *Pius XII, Karol Wojtyla and Joseph Ratzinger* through political means. The result can be equally devastating, millions dead in war versus a ban on contraception that results in millions more dead of AIDS and mass poverty and starvation in third world countries.

The end of a dream

Nevertheless, had Hitler won the war, everything Karol Wojtyla had ever dreamed of would have become a reality. We would today be living in a world of white Aryan male superiority rather than in this world in which we find ourselves, this world of equal human rights and dignity for everyone, this anti-*Christian,* anti-*Fascist* world of fairness and justice for all which, as we speak, threatens to be extended to homosexuals.

However, when Hitler lost the war, much of what Karol dreamed of was taken from him as the door had been opened to move civilization away from *Fascism,* away from Moses, toward Christ - toward the equality of all men and women.

Again, in criticizing John Paul for clinging to *Fascism,* one must consider his upbringing. Any psychologist will tell you the most formidable years are ages six through fourteen. During those years, Karol was dressed in the military uniform of a *Fascist* school and was guided by two devout *Fascist* parents. One must also understand his successor, Joseph Ratzinger (Benedict XVI); for he too, during his most formidable years - ages six through fourteen - was guided by two devout *Fascist* parents and wore the military uniform of the *Hitler Scouts.* The handbook of the *Hitler Scouts* was *Mein Kampf.*

One must be understanding of these popes, unable to break free of the roots of *Fascism* they grew up in, they continued to cling to fragments of *Fascism* society has not yet taken from them.

However, one must be fair in these things. Homosexuals have the misconception John Paul II did not accept them for who they were. From the day he took office, John Paul accepted the ruling of the world psychiatric community that homosexuality was a matter of human nature. This is seen clearly in that he discouraged sexual orientation change therapy in the Church.

What he did not condone was sex outside marriage. His chastisement of homosexuals has been no more severe than has his deprivation of equal human rights for the remarried and others living outside of what he considers marriage. Both John Paul II and Benedict XVI do not believe the *falling in love* syndrome defines marriage. They decree that two people of the right body parts constitute marriage whether or not they intend to parent children.

Lucien Gregoire

Albino Luciani 1942

Albino Luciani, on the other hand, was born to very different parents; a devout mother who gave him the imagery of Christ and a social revolutionary activist father who gave him the reality of Christ. As was the case in Poland, blacks were not allowed in Italy while he was growing up.

His father objected to discriminatory practices. He taught the boy Luciani from an early age all of God's children were created equal regardless of sex, race, creed, sexual orientation, physical or emotional impairment or what have you. Unlike Wojtyla and Ratzinger, Luciani did not believe in what Pius XII and Hitler had to offer. When Hitler lost the war it was the beginning of the realization of all he had ever dreamed of.

Luciani is pictured above wearing the soft cap of the *Italian Resistance*. During the war he developed a knack for mountain climbing when he ministered the members of the Italian Resistance who were hiding in difficult to get at places in the Dolomite Mountains. He once told me of the most difficult moment of his life; he felt the pulse of a young boy run out between his fingers. The boy had been shot trying to blowup a railroad trestle to cut off military supplies coming from Germany.

White Light Dark Night

The ecclesiastical record of two popes

	Luciani	Wojtyla
Vow of poverty (all clergy)	yes	no
Planned Parenthood – contraception	yes	no
Ordination of women	yes	no
Homosexuality is God's will	yes	yes
Love and commitment define marriage	yes	no
Sex defines marriage	no	yes
Sanctification of remarriage	yes	no
Black-white integration	yes	no
Genetic research	yes	no
Mysticism	no	yes

In the twenty years they were bishops, Luciani and Wojtyla were among the loudest voices on their respective sides of the aisle, the reason why they had both risen to the papacy. For example, after the war, Wojtyla rose as the leader in the effort to preserve segregation between blacks and whites within the Church, and Luciani rose as the leader of the movement in Europe to do away with segregation, not only within the Church but in society as well. One must keep in mind that, at the time, the population overwhelmingly favored segregation.

"Children of the Devil"

Most Catholics are of the misconception a pope is restricted by scripture or the actions of his predecessors in making changes to Church doctrine. Nothing could be further from the truth, as a pope by *Canon Law*, has the absolute power to unilaterally make any change he deems fit.

Just before Luciani was raised to the papacy, he had declared Louise Brown, the first artificially inseminated child, *". . . a child of God,"* which contradicted the papal decree made thirty years earlier by Pius XII, *"Such children if ever born would be children of the Devil."* Because Luciani had shortly risen to the papacy, Louise Brown was recognized by Catholics worldwide as being a child of God.

On the day following his election, John Paul II clarified the confusion caused by his predecessor's action. He issued a decree reaffirming the decree of Pius XII and he did it in such a way as to question the possible ensoulment of artificially inseminated children. In 1984, public pressure forced him to reverse this doctrine - his own doctrine of just six years before - a change that today recognizes the ensoulment of all children, whether naturally or artificially conceived. So although one might think popes infallible, there have been times popes have not only reversed doctrine made by a predecessor, but have often reversed doctrine made by themselves; and this is one of them.

Contraception and Abortion

When one believes a pope cannot ordain women, cannot accept homosexuals into the Church, cannot sanctify remarriage, and cannot take a vow of poverty - one does not know what one is talking about. For the Pope is an absolute dictator of not only the present, but of the past, and of the future.

Popes who impose a ban on contraception, the driving force behind poverty, starvation and the spread of AIDS in third world countries, are relying upon more than simply Church doctrine. The theological basis for the Church's position is *'Of the thousands of sperm and eggs that a man and a woman produce in their lifetimes, God has pre-selected that particular sperm and egg that will become a human being.'* Contraception, stem-cell research, abortion of a fertilized egg, and even masturbation interfere with this natural flow of creation and are therefore mortal sins. It is this thinking that causes the preacher to consider a fertilized egg to be a human being; the reason why *Canon Law* considers all sex other than copulation within marriage to be a mortal sin as it interferes in God's selection of the parts that form a human being.

This ideology - God has pre-selected that particular sperm and egg that will be a human being - is supported by *Jeremiah 1,* "The Lord came unto me saying, Before I formed thee in the belly I knew thee; and before thou camest forth out of the womb I sanctified thee." Take this line out of the Bible and Christianity has nothing at all in scripture to support its ban on contraception, stem-cell research and abortion of a fertilized egg. [1]

Conversely, Moses' testimony, the only condemnation of abortion in the Bible, explicitly contradicts Jeremiah. *Exodus 21, "If men strive and hurt a woman with child, so that her fruit depart from her, and yet no mischief follow; he shall be surely punished, according as the woman's husband will lay upon him; and he shall pay a fine as the judges determine. Yet, if any mischief follow, that thou shalt give life for life, eye for eye, tooth for tooth, hand for hand, foot for foot."*

According to the 'God' Moses spoke to, abortion of a fertilized egg is a misdemeanor while abortion of a viable fetus is murder. According to the 'God' Jeremiah spoke to, abortion of a fertilized egg is murder. To take it a step further, destruction of a viable sperm or egg before fertilization is murder; the reason why the Church views masturbation as mortal sin. It depends entirely on which prophet one thinks was telling the truth.

It is this concept that divides pro-choice from pro-life: pro-choice being against the destruction of a viable fetus, while pro-life being against the destruction of a fertilized egg which is not yet an embryo; despite the fact that billions of them are flushed down the sewer in everyday sex. In the United States, it is that the preacher thinks a fertilized egg is a human being that causes him to ban the morning-after pill which results in a million avoidable abortions or murders of fetuses each year; something pro-choice people are trying to prevent.

Pro-life protects fertilized egg   Pro-choice protects viable fetus

It is this verse in *Jeremiah* that supports the premise that ensoulment takes place just before birth, the reason why no church baptizes before birth; also the reason why preachers are against aborting malformed or brain-dead fetuses. It makes no difference to the preacher whether or not a child is born only to suffer and die an unspeakable death. All that is important to him is that the child reaches that point at which it acquires a soul and goes to heaven.

In 1984, when public pressure forced John Paul II to reverse his own doctrine of 1978, which had denied ensoulment to artificially inseminated children, he took exception to his own basic premise as, in the case of artificial insemination, man and not God selects which sperm and which egg will become a human being. In the process of selection man destroys many sperm and eggs that are found to be disease or genetically contaminated.

Today, the Vatican's position concerning artificial insemination contradicts its most fundamental dogma of human theology.

The twelve year old child prodigy Albino Luciani of the Feltre Seminary had once published in his school paper, *"The great enemy of society is not so much preachers driven by ego, greed and bigotry who preach these things, but it is the ignorance and the weakness of the minds of those who are fooled by them"* [2]

Luciani was to live to see his supposition proved true. In 1946, he told a reporter, *"It was not so much Hitler and Mussolini who were the culprits in this thing, as it was the ignorance and the weakness of the minds of the masses that believed in them"* [3]

Unfortunately, as we shall see in what is yet to come, the papacy remains today the last remnants of the *Anarchy of Fascism* the teenager Albino Luciani once warned of in his editorial that criticized the dictate of Pius XI to enroll all children under the age of sixteen in the *Italian Fascist Organization* which eventually cost fifty million lives. The very same *Anarchy of Fascism* that is about to cost fifty million more in the spread of AIDS in unprotected sex in third world countries; *"the weakness of the minds of men."*

Two very different Gods

During the twenty years they served as a bishop and as a cardinal, each of their countries suffered from an immense orphan problem - about two million in each country. During that time Wojtyla built and dedicated fifty-three churches and not a single orphanage. Luciani, on the other hand, built and dedicated forty-four orphanages and not a single church.

Each time the fork in the road would come up, Wojtyla would ask himself, *"Now, what would the God of Moses do in this case?"* Each time the fork in the road would come up, Luciani would ask himself, *"Now, what would Jesus do in this case?"*

So these were two very different kinds of men, driven by two different kinds of Gods. Karol Wojtyla striving to preserve the last bits and pieces of *Fascism*, the law of the *God of Moses*, and Luciani striving to bring the reality of Christ into a modern world; striving to stamp out what Hitler stood for, once and for all!

The Imitation of Christ   **The Imitation of Moses**

John Paul I   **John Paul II**

[1] pg 138 Because *Isaiah* spoke of an entirely different God than the God Moses spoke to, 'experts', through the centuries, have tried unsuccessfully to link the *God of Isaiah* to the *God of Moses*. Yet, concerning every major issue the two Gods differ. For example, the *God of Moses* treats sex as the greatest of sins; whereas the *God of Isaiah* scarcely mentions it and in no case condemns it. Christ differs from the *God of Moses* on major issues, yet agrees with the *God of Isaiah* on major issues.
[2] pg 140 *Parish Bulletin Feltre* 16 May 25
[3] pg 140 *Messaggero Mestre* 1 Apr 46

Chapter 9

How a Pope is Elected

According to *Canon Law*, the decision as to who becomes Pope must be Christ's decision and not the decision of a group of men who are collaborating among each other. This is why after a candidate achieves the minimum margin of victory a revote is taken to make it unanimous; Christ has spoken, so to speak.

The public perceives the electors as being a group of men cloistered together in the Sistine Chapel in prayer. It believes the only guidance a cardinal has in making his choice is prayer. As a matter-of-fact, prayer has very little to do with the election of a pope.

Like any other election, the election of the Supreme Pontiff of the Roman Catholic Church is a political process. A cardinal does not cast his vote for another cardinal because he thinks he will look pretty in a white satin gown. Likewise, he does not cast his vote for a man because Christ appears to him the evening before and tells him to. Like any other republican or democrat, the voting cardinal casts his vote for the individual who most closely shares his own ecclesiastical ideals. As in any other election, he must often trade off a part of his own convictions in order in return to get a piece of the action.

In modern times, with the exception of John Paul I, popes have handpicked their successors. Historically, the reigning Pontiff would move his choice into the Secretary of State position or into the primary pastoral archdioceses of Milan or Venice.

The majority of twentieth century popes have come out of these three jobs. This included Pius XII who served as Pius XI's Secretary of State. It included John Paul I, John XXIII, Paul VI and others.

John XXIII and Paul VI

When Pius XII advanced in age, he realized the popular Archbishop of Milan, Giovanni Montini, would most likely get the vote. Montini had gained immense popularity with field cardinals when, as Undersecretary of State in the Vatican, he had often embarrassed Pius publicly on humane issues. In the early fifties, Pius transferred Montini to Milan under the guise of a promotion in order to remove his liberal influence from the *Vatican Curia*.

Pius took two strategic steps to prevent Montini from succeeding to the papacy. In 1953, he moved his old conservative crony Angelo Roncalli into the coveted position of Patriarch of Venice and at the same time made him a cardinal.

On the other hand, Pius refused the *red hat* to Montini, despite the fact that Milan was by far the largest congregation in Italy. The Pope's intent was to exclude Montini from the conclave which would greatly diminish his chances of being considered. It worked. Five years later when Pius died, Montini, not being a cardinal, was excluded from the conclave; Roncalli got the vote.

Roncalli (John XXIII), on the other hand, in his first official act, named Montini a cardinal and during his brief reign he often publicly referred to the Archbishop of Milan as his chosen successor. To the most casual of onlookers, this was in return for Montini's influence with the conclave cardinals who had made him Pope. When John died in 1963, Montini stepped in - Paul VI.

Benedict XVI

In that popes name their successors is perhaps no better exemplified as was the case of John Paul II. Throughout his papacy, he had purged the *College of Cardinals* of liberalism. In addition, he carefully purged the voting cardinals of those who might not vote for his choice.

Of the 115 voting cardinals at the time of his death, only two had not been named a cardinal by John Paul II. He could be reasonably certain that a conservative would succeed him. Yet, he could not be certain that the *College* would select his personal choice, his best friend and closest confidant, Joseph Ratzinger.

The Dean of the *College of Cardinals* had historically been an honorary title that was conferred upon the oldest living cardinal, simply a figurehead position. It had no responsibilities or, for that matter, any authority. In November of 1998, John Paul II elevated the position to one of the most powerful voices in the Roman Catholic Church, giving it broad doctoral powers. He arranged a closed election for the new position among the conclave cardinals and confirmed the results of that election in November of 2002.

The winner of that vote was Joseph Ratzinger. When John Paul II went to rest, he among all popes was certain that his choice would succeed him. In that, the very same group of cardinals had elected a man in a secret election to be the ranking cardinal, the second most powerful position in the Church, they would surely elect him to the top job in a successive vote.

Yet, there was no case more widely known than was Paul VI's intent that Albino Luciani would succeed him.

Shoes of a Fisherman

Of all the demands Luciani had made upon the Vatican through the years, none was more widely publicized than was his ongoing request the Vatican liquidate its treasures to annihilate poverty and starvation. In his early years, Luciani didn't have to look further than his own home town to see poverty and starvation. In the nineteen-sixties, as a bishop, he saw much more. He visited impoverished areas of Africa and by 1966 half of the clergy of the Veneto country was working at least part time in Africa.

Luciani, on an ongoing basis, brought public pressure on the Vatican to sell those treasures held in warehouses and not on display to help finance his efforts, but to no avail. Yet, Paul VI, himself, did sell many of his personal belongings and sent the proceeds to Luciani. Nevertheless, of all the things Luciani had gained notice for, was his condemnation of the hypocrisy of the Vatican treasures.

It was during this time that he struck up an endearing and lifelong relationship with Cardinal Gantin, the Primate of Africa, and his counterpart in the eastern world, Cardinal Yu Pin, who at the time was Primate of China; Africa and China being the most widespread areas of poverty in the world.

In the fall of 1968, the Anthony Quinn movie *Shoes of the Fisherman* previewed. The film capitalized on the reputation of the popular bishop from Vittorio Veneto. It depicted the rise of a Russian to the papacy who resembled to a great extent the real life liberal Luciani to a tee. It told of a bishop who would often walk in common street clothes incognito through the darkened ghettos of his diocese under assumed names and, perhaps, most pointedly of all, the film exploited Luciani's most widely known threat to Rome - his ongoing demand to liquidate the Vatican treasures to annihilate poverty and starvation in the world.

In the film, Anthony Quinn, a newly appointed pontiff, shocks the world by announcing his intent to sell off the Vatican treasures to annihilate poverty and starvation in Communist China.

That the leading character in the film so closely resembled the real life Luciani, and for the most part ignored the character that Morris West had built into his 1963 novel, brought the bishop of Vittorio Veneto much notoriety. For example, there is no mention of the liquidation of Vatican treasures in the book at all, which is the focal point of the movie.

Nevertheless, the movie was a Godsend for Luciani and in the coming months the fame it brought him would finance many of his African and Asian ventures.

The year following the film's release, Paul VI raised Luciani to Archbishop of Venice. Together with the appointment, Paul sent a public message to the bishop of the remote mountain province. *"The time has come for you to begin your journey to Rome, for the gods of Hollywood have spoken!"*[1] Everyone in Europe knew exactly what Paul meant by his remark.

Whenever, various prelates gathered together in Rome and Luciani was among them, though only a simple bishop, he was always seated next to Paul. In fact, in one well publicized instance at a public audience, Paul could not locate the little buzzer on his chair that would summon an attendant. Luciani reached for Paul's hand and guided it to the button. Paul looked at him and, not aware the microphone would pick up his comment, said, *"So, you already know where the Papal buzzer is."*

In 1967, at Christmas service in Vittorio Veneto, Pope Paul removed his Papal stole and placed it on Luciani's shoulders. During his pontificate, Paul often repeated this gesture. Yet, it usually went unnoticed until he did it in 1972 in the Piazza San Marco in Venice before twenty-thousand people and an international television audience. The incident was reported in the world press. Now the whole world knew Luciani was the choice.

In the seventies, as Paul was aging, it became obvious that several of the changes he was making were designed to assure the election of his liberal friend from Venice in a conservative Church. He began stacking the *College of Cardinals* with liberals. In 1973, he raised the number of authorized cardinals from seventy to one hundred twenty and filled most of the vacancies with liberals. Of the last thirty-eight cardinals he named, thirty-four were liberals.

Soon afterwards he made the change that eliminated those cardinals over eighty from voting; the reasoning being that aging cardinals tend to lose their touch with the world and are too set in their thinking and therefore should not be allowed to vote. Yet, in retrospect, this made no sense as Paul, himself, was approaching eighty. To exclude those over eighty from the voting process he would be admitting that he, too, had lost touch with the world. He made this change to remove eighteen conservative cardinals from the voting conclave, as all those over eighty were conservatives.

As a follow-up to having raised the voting cardinals to one hundred twenty, he raised the authorized number of cardinals to one hundred thirty-eight in order to hold the voting conclave at one hundred twenty. This created eighteen additional voting vacancies, nine of which had, in rapid consecutive order, been filled by Paul with liberals before his death. Had Paul not made these changes, his successor would not have been Albino Luciani.

When he lay on his death bed, Paul could be reasonably confident through the simple process of addition that Luciani had about seventy-five votes; a marginal win. Yet, he could not be certain his choice would succeed him.

Barely conscious, falling in and out of a coma, on the night before his death, Paul promoted Luciani's great friend and supporter, Cardinal Yu Pin of Taiwan, to the position of Grand Chancellor of Eastern Affairs; the most powerful position in the Eastern Hemisphere. Paul's intent was to give Yu Pin the influence he would need in the upcoming conclave to add the dozen or so eastern cardinals to Luciani's list. When Paul went to rest, he knew that Albino Luciani, the brave young priest he had once saved from excommunication in the wake of Pius XII, would succeed him.

The Opus Dei candidate

Although he rarely traveled outside of Italy, at the time of his election, Luciani was the most widely known cardinal in the world.

On the other hand, his successor Karol Wojtyla was scarcely known outside of Poland, though he was the most widely traveled cardinal in the history of the Church.

Wojtyla, being an ultraconservative, brought nothing new and rarely made the headlines. Conversely, as an outspoken ultraliberal, Luciani was constantly in the press. Today few of us can name very many cardinals as they are mostly conservatives and rarely make headlines as the liberal Luciani often did.

Nevertheless, all modern popes have acceded to the papacy through the influence of their predecessor; with the exception of Karol Wojtyla. One might ask how it was possible that this non-Italian bishop, practically unknown in the media, wiggled his way into the papacy.

Being the ultraconservative he was, Karol knew he would never get the nod from Paul VI, a moderate liberal. Knowing this, he knew his only chance was to go to the voters themselves. We all know him as having been the most widely traveled pope in history. We are less aware he was also the most widely traveled cardinal in history. In the ten years he was a cardinal, Karol visited more than one hundred fifty cities around the globe. He visited literally every single one of the field cardinals, including his well publicized trips to the United States in 1969 and 1977 in which he visited every city in North America where a cardinal was in residence.

There is no record in the press that any of these trips had anything to do with fund raising. One might ask, what other reason could he have had to have spent so much money from the poor box for so many expensive vacations? What could these excursions have possibly had to do with his responsibilities as Archbishop of Krakow? The only logical answer is he was lining up the votes.

In truth, Karol did not rob the poor box to pay for his trips. Like Albino Luciani, Karol Wojtyla had a benefactor who would pave his way to the top. The Polish cardinal's ten year 'campaign' was funded by his good friend Josemaria Escriva, founder of Opus Dei; Wojtyla was the Opus Dei candidate to succeed Paul VI.

Wojtyla and Escriva met in 1965 at the Vatican II Council in Rome. The next year Karol made his first campaign trip. The following year, through the influence of Escriva's closest friend, Agostino Casaroli, Wojtyla was named a cardinal.

From an economic point of view Opus Dei is a commune; it requires its members to contribute the bulk of their income to the organization. It is no secret Opus Dei uses its vast financial resources for political purposes. It is no surprise it financed the Krakow Archbishop to the pinnacle of the Catholic world.

Shortly after being raised to the papacy, John Paul II made Agostino Casaroli, a common bishop, a cardinal and raised him past four hundred others who outranked him to *Secretary of State,* the second most powerful position in the Church. Casaroli was, at the time, one of the most influential members of Opus Dei. In 1982, in the face of raging controversy, Wojtyla made Opus Dei a *Prelature of the Holy See,* changing its reporting status directly to the papacy. In 1992, when Casaroli retired, Wojtyla replaced him with Angelo Sodano, also a powerful member of Opus Dei. In 2002, again in the face of raging controversy, Wojtyla elevated Josemaria Escriva to sainthood. In their silence, two-thirds of the world's bishops objected to both the *Prelature* and the canonization. Jewish and many other communities were infuriated.

Opus Dei is an order of *right wing* extremists. It believes salvation can be obtained solely through prayer and adoration. It does not believe helping others has anything to do with it. Although Opus Dei is, on a per capita basis, one of the richest organizations in the world, it does little to help others. Its sole brush with charity is its *Harambee* mission in Africa which purpose is to indoctrinate youth into its fascist fold.

Opus Dei was founded in 1928 when movements both within and outside the Church started to move Catholicism toward the *left.* Jesuit priests, in defiance of Canon Law, began to baptize born-out-of wedlock children. Women gained the right to vote and were threatening to leave the kitchen. For the first time, Jews were permitted to practice their religion in Catholic countries. In Europe, homosexuality was gaining some level of tolerance. On the world stage, the Soviet Union, the great enemy of fascism, was coming to power. In his own country of Spain, democracy was raising its ugly brow. All these things horrified Escriva.

The handbook for his new fascist order was *The Way.* The core of its message for salvation, *"Blessed be pain. Loved be pain. Sanctified be pain. Glorified be pain."* [2] For some parts of his book, Escriva was accused of plagiarizing Hitler's *Mein Kampf.*

The accusations were not well founded; it was just that Escriva and Hitler shared the same ideology.

When the Franco regime came to power in 1939, Escriva came immediately to the ruthless dictator's side. Opus Dei members rapidly infiltrated the cabinet and brought about oppression of the Spanish people which included a resurgence of the Inquisitions; something Opus Dei believed in.

It cannot be proved that Opus Dei in itself was a killer. Yet, it was so entwined in the Franco anarchy that they were one and the same body. Franco provided the hit men and Opus Dei provided the ideology. Together they murdered over a million innocent people.

Yet, there are some deaths directly attributed to the Catholic cult itself that have made their way into books and motion pictures. One might try the film, *"The Disappearance of Garcia Lorca"* for a starter. There are many others.

In July 1941, Escriva praised his greatest ally, *"Hitler will take care of the Jews. Hitler will take care of the Slavs."*[3] After the war, Escriva minimized the damage, *"History is unfair to Hitler. It claims he murdered more than six million in his concentration camps, whereas, less than four million died."*[4]

In 1946, he enlisted Carlos Fuldner into the ranks of Opus Dei. Fuldner had been an officer in Hitler's SS Guard and an old friend of Escriva - probably why Escriva was able to make his remark in 1941, *"Hitler will take care of the Jews..."* as only the SS Guard knew what was going on in the concentration camps at that early time. Fuldner was assigned the job to run rescue efforts from Madrid to Argentina and Brazil for Nazi criminals seeking refuge. Two of the more famous of those rescued by Opus Dei were Adolph Eichman and Josef Mengele.[5]

In early 1978, the dimmest insider in the Church knew Luciani controlled a substantial block of votes in the next election. The Primate of Opus Dei, Alvaro de Portillo, sent a short note to Luciani asking him to be the keynote speaker at its upcoming convention. The strategy was to win Luciani's support for Wojtyla in the next election. In his invitation, Portillo cited the commonality of Luciani and Escriva; *"the Imitation of Christ."*

Luciani declined. He told Portillo, *"True. Msgr. Escriva and I are in common. We both believe the Imitation of Christ is the path to holiness. Yet, Escriva's teachings are radical and materialistic. He believes salvation can be had solely through the Imitation of Christ's death, self-flagellation and the chanting of vain repetitions. I believe salvation can only be had only through the Imitation of Christ's life, in helping others . . .* "[6]

When the convention met the following month, Portillo introduced Opus Dei's candidate as *"Papa Stanislao."* The crowd broke into a frenzy for thirty minutes before Karol Wojtyla was able to begin his speech. It was commonly known that *Stanislao* was Karol's chosen name should he ever rise to the papacy.

Six months later when Luciani won the election, Villot, the *Secretary of State*, was eighty and there was growing expectation he might retire. If that were to happen, the shoe-in for the most powerful administrative position in the Roman Catholic Church was Luciani's good friend, Giovanni Benelli. Benelli was the outspoken enemy of Alvaro de Portillo and Opus Dei. He would bring a swift end to them. When Luciani won the election, Opus Dei was as good as dead. If one rationalizes it was Opus Dei that orchestrated the murder of John Paul I and those around him any court in the world would justify it as self-defense.

In fact, there was wide speculation Benelli had surrendered the block of votes he controlled to Luciani – which Luciani needed to lock up the election - in exchange for the No. 2 spot on the ticket. It is more likely Benelli gave them to Luciani anyway, as the two shared the same ecclesiastical objectives and a common enemy, Opus Dei.

When Wojtyla rose to the papacy, like Franco before him, much of his cabinet was infiltrated by Opus Dei members. Even his personal secretary Father Stanislaw Dziwisz, was an avid cult member as if to hint His Holiness himself required an occasional flogging of the rump while he prayed the rosary.

When, in 1982, he raised Opus Dei to *Prelature of the Holy See,* he did it in the face of an uproar, not only among Jews, Slavs, homosexuals and other targeted groups, but among men and women of good conscience all over the world. He did it at a time he, himself, was caught up in the middle of the *Great Vatican Bank Scandal,* something that would have resulted in impeachment of any other head of state.

The mysterious death of Cardinal Benelli

After the assassination attempt on John Paul II in 1981, Joseph Ratzinger was moved to Rome as *Prefect of the Congregation of the Faith.* The move positioned Ratzinger to replace Wojtyla should anything happen to him. Of the cardinals, Ratzinger most closely shared Moses/Hitler/ Franco/Escriva/Opus Dei ideology.

Another event precipitated the *Prelature.* Banco Ambrosiano collapsed on August 6, 1982 shortly after its president Roberto Calvi was found swinging from Blackfriar's Bridge. This caused the courts to focus on senior figures of Opus Dei who were believed to have lured Calvi to London on the guise of a loan to get him out of his predicament. Calvi and his secretary, who was found dead in Milan on the same day, were the only people who could have known of Opus Dei involvement in the bank scandal.

The Vatican's announcement of the *Prelature* on August 23, 1982 claimed John Paul II had made the declaration on August 5th. As a *Prelature of the Holy See,* Opus Dei officers were immune to Italian Courts. The timing of the Prelature to the day before the bank's collapse is powerful evidence that Opus Dei had been involved in the fraudulent deception of European investors - *the Great Vatican Bank Scandal* - and the murders that surrounded it.

Knowing few cardinals favored the *Prelature,* Cardinal Benelli called for its ratification by the *College of Cardinals* setting the date for Nov 1st. Although the Pope had absolute authority, Benelli would force him to reconsider. Benelli, in robust health, was in the middle of lining up the opposition when he suffered a heart attack.

There are lethal toxins that can precipitate heart attacks in healthy people. In its last release, the Associated Press reported Dr. Antonini as reporting Benelli had been placed on dialysis and had *"shown positive signs of recovery."* Rumors of foul play flourished as toxins attack the purification organs, particularly the kidneys. The next day, an order came from the Vatican to remove him from dialysis and send him to his home where he died a few hours later.

The Vatican, to this day, insists Benelli, himself, asked to be removed from life support; that is, committed suicide. This is to say that the greatest enemy of Opus Dei had on the eve of his greatest victory, committed suicide. His enemy Agostino Casaroli flew to Florence and took control of the body, said a funeral mass and buried him beneath the basilica. In 1982, mass and consecrated ground were forbidden for suicide victims by Canon Law which demonstrates the Vatican knew he hadn't killed himself. Needless to say, the vote to deny the *Prelature* never took place.

Nevertheless, when the first conclave opened, Wojtyla was the only member who had a personal relationship with each of the others. He had slept in each of their mansions, had tasted wine with them at dinner and enjoyed morning breakfast with them on their verandas. To believe that not one of them voted for him in the first conclave of 1978 is ludicrous. To say that he went from zero to two-thirds of the vote in a single ballot in the second conclave is every bit as ludicrous. Particularly ludicrous, as within conclave rules, a cardinal can only turn to prayer for guidance.

Speaking of conclaves, unlike what the Vatican, the press and motion pictures might lead one to believe, popes are not elected in conclaves. One can see this clearly if one considers the conclave rules of secrecy as they apply both before, during and after a conclave.

If a cardinal discusses publicly anything relative to an election before, during or after an election he automatically self-excommunicates himself. This is confirmed in that no cardinal has ever talked publicly about anything that pertains to an election whether it occurred before, during or after a conclave.

Under the conclave rules, cardinals are permitted to nominate and discuss candidates before an election privately among themselves. In the two week period between a pope's funeral and the next conclave, many of the voting cardinals are sequestered together in the Vatican. Those that are not physically present are no further away than is the nearest telephone. Unlike what the public is led to believe, most of this time is not spent in prayer, it is spent electing the next pope. There is nothing in the rules forbidding nominating, lobbying, negotiating or tallying of votes before the conclave begins, provided it is done privately.

In retrospect today, in that popes in modern history have been elected on either the first or second day, it is obvious the cardinals have already known the choice, or at the very least the leading choices, when the conclaves began.

Otherwise, in a secret election, as required by the conclave rules, it would take years before anyone would by chance come up with two-thirds of the votes. Yet, one knows from the final counts in recent elections, the frontrunner entered each of those elections with a marginal number of votes as they each won by the minimum margin required.

Once the conclave convenes, the rules tighten up. Unlike Hollywood might portray, no nominating, lobbying or negotiating of candidates is permitted. The only lobbying permitted is silent prayer. Each cardinal writes a name on a small square of paper and folds it and places it into a large chalice. Three cardinals, who have been chosen as counters, count the ballots in a private room. If they do not come up with a winner, they burn the ballots and a release of black smoke appears from the Sistine Chapel signifying to the public the vote has failed. According to *Canon Law*, the actual count of each ballot is not announced to the conclave.

One might wonder how the voting cardinals know who the leading vote-getters are at a given point. As just pointed out, they most likely know when the conclave begins. There is a supposition that is based on innocent Vatican releases that have reported that the winner in modern elections has been consistently seated in the center chair of the first row on the St. Peter's side of the Chapel. These same press releases through the years have also confirmed that the most likely runner-up was seated in the center chair of the first row on the Apostolic Palace side of the Chapel. This would suggest that at the end of each ballot, the counters place the leading vote getters in a pre-arranged seating arrangement.

For example, in the election that chose Montini in 1963, according to the *London Times*, *"... The Secretary of State approached the cardinal from Milan who was seated in the center of the first row and directly opposite him was Cardinal Siri of Genoa..."* and in the case of Luciani's election, *"... Cardinal Villot came to the cardinal bishop in the center seat. Directly opposite him was seated Karol Wojtyla of Poland..."* and again in Wojtyla's election, *"... Villot placed his hand on the shoulder of the Polish cardinal who was seated in the center chair..."*

This is seen more clearly when one considers changes that John Paul II made to the conclave rules in 1995. Realizing that frontrunners had entered recent conclaves with marginal numbers of votes which could have gone either way and possibly have stalemated an election, John Paul added a stipulation that if the conclave fails to agree on a choice within so many days, and so many days after that, the cardinals will break for a day, here and there, for 'prayer' - the same kind of 'prayer' that goes on during the two weeks before the conclave opens.

This gives the cardinals the freedom in future elections to negotiate outside of conclave rules if an election were to stalemate. John Paul made another change which greatly diminished the influence of the counters in the conclave. He lifted the secrecy of the vote. That is, in future elections each cardinal, in addition to writing the name on a piece a paper and placing it into the chalice, will announce the name of the cardinal he votes for.

This still leaves the possibility that a cardinal might call out a name other than that which he puts in the chalice. Nevertheless, the point is that a papal election is just like any other election, the winner is determined before the count begins.

In the case of popes who are ailing for sometime before death, the lobbying and tallying goes on in the months before the cardinals arrive in Rome. This was true of John Paul II and it was also true of Paul VI.

When the cardinals first gathered around Paul's coffin, each of them knew Luciani had at least eighty votes; more than enough to win on the first ballot. Each of them knew that Paul, the evening before his death, had appointed Yu Pin as *Grand Chancellor of Eastern Affairs* which put the Venice cardinal over the top. Each of them knew that if something where to happen to Yu Pin, Luciani's chances would be seriously diminished.

Media mayhem

There is no other major world event in which the public is more vulnerable to the press than is a papal election. The reason is that no one other than the conclave cardinals knows what goes on in a conclave. Everyone has to guess and this includes the press. This affords the press the opportunity to sensationalize events.

Despite the fact Paul had named Luciani his successor and had made many strategic moves publicly to that end, no newspaper in the world considered him a candidate. Despite the fact the *London*

Times knew Wojtyla had campaigned around the globe for the votes and had itself reported that Wojtyla was seated in the *runner-up chair* in the Luciani election, it did not consider him to be a factor in the second conclave of 1978. Instead, it predicted a two-way race between Benelli of Florence and Siri of Genoa.

Despite the fact the *Times* had missed the boat entirely, as it had published a list of a dozen leading candidates led by Benelli and Siri in each election and not listed either Luciani or Wojtyla among them; the day after Wojtyla was elected, it issued the statement, *"Unable to decide upon an Italian cardinal on the first day, the College decided to look elsewhere and elected the Polish cardinal on the second day."*

Everyone believes this to this day despite the fact that no one other than the participating cardinals knows what actually took place. Everyone believes this to this day, despite the fact that the media had demonstrated quite clearly in its many predictions that it knows nothing about what goes on when a pope is elected.

This is how books and motion pictures have described the election of John Paul II ever since - *"Unable to decide upon an Italian cardinal on the first day, the College decided to look elsewhere and elected the Polish cardinal on the second day."* As if to say, a man who had not gained a single vote on the first day, in a closed election in which the only lobbying and negotiating permitted is prayer, gained a two-thirds majority on the morning of the second day. Karol Wojtyla, like his predecessor Luciani, already had most of the votes he needed to win when the conclave that elected him opened on the first day. What's more, he knew it.

[1] pg 145 L'Osservatore 6 Mar 73
[2] pg 148 The Way 208.
[3] pg 149 Metro Madrid 2 Mar 41
[4] pg 149 IL Mondo 7 May 47
[5] pg 149 Odessa File Frederick Forsyth
[6] pg 150 Messaggero Mestre 3 Feb 78
[7] pg 152 In 2005, five members of the Sicilian Mafia were brought to trial charged with the murder of Roberto Calvi. Two years later in June 2007, all five were acquitted. There were only two organizations with motive to kill Calvi, the Mafia and Opus Dei. With Diplomatic Immunity, Opus Dei has never been brought to trial, one reason it remains a Personal Prelature of the Holy See.

Chapter 10

Strange Events of August 1978

There were some events that surrounded the funeral of Paul VI that may have been related to the first conclave of 1978.

The day before Paul's funeral, Belgium radio announced Cardinal Suenens, Primate of Belgium, had been killed by a falling section of an aging building façade in Brussels. The radio report, which was based on eyewitnesses of the event, was premature.[1]

It had been a visiting French bishop that had been killed. It had been that the incident had occurred near the cardinal's palace and that the bishop had been wearing black garb topped off with a red zucchetto that caused witnesses to mistake him for the cardinal.

Of course, if this was true, that witnesses on the ground made such a misjudgment, then it would have been likely that anyone on the roof of the building would have made the same mistake. Suenens, of course, being the recognized leader of liberalism in the Catholic world at the time would be one of the most influential cardinals in the upcoming conclave.

As already mentioned, realizing death was imminent, Pope Paul elevated Cardinal Yu Pin to *Grand Chancellor of Eastern Affairs,* making him the ranking and most influential cardinal in the Eastern Hemisphere.[2] One can only surmise that Paul's motive, knowing that Luciani controlled a very marginal number of votes to win the upcoming election, was to increase Yu Pin's influence in the conclave and thereby lock up the election for his chosen successor. One can only wonder why else was this appointment so important that Paul, on the day before he died, lapsing in and out of a coma and barely able to speak, would make such an effort.

Regardless of the appointment, Yu Pin would have been the most influential member of the conclave from the east anyway. It had been through his influence with Paul that a half-dozen eastern bishops had been made cardinals; though conservatives, they would vote for Yu Pin's good friend Luciani because they owed him one. True, the appointment may have increased his influence with other eastern cardinals and gathered a few more votes, but Yu Pin already had a half dozen locked up anyway. Of course, if anything were to happen to him, even that half dozen would no longer owe him one.

Yu Pin would not learn of his step up in the eastern world until after Paul had died. At the time he was promoted, he was in the air returning to Taipei from Venice where he had been visiting Luciani. In fact, when he arrived in Taiwan, he learned of Paul's death and boarded another plane and flew back to Rome.

Four days later, Cardinal Yu Pin keeled over at the funeral of Paul VI. The Vatican cited a heart attack despite the fact Yu Pin had no history of heart disease. The press, suspecting poison, demanded an autopsy. Cardinal Benelli of Florence followed with a formal request for an autopsy. As had been its policy in other unexplained deaths, the Vatican refused an autopsy. The body was embalmed and returned in a sealed coffin to Taipei for interment.

A week before the election came another strange incident. This one involving Cardinal Benelli. He barely escaped death when a small section of a frieze fell from a Vatican building narrowly missing him and his secretary by inches. Although the Vatican buildings are aging, this is rare, as the facades are routinely checked for defects to protect the tourists that roam Vatican City. [3]

The incident carried small notice in the Vatican newspaper the next day calling for increased inspection of the buildings. In Brussels, the situation was different. Because the incident resembled so closely that which had killed the bishop in Brussels two weeks earlier, the press began to point fingers. Yet, in time, it passed the incidents off as having been mere coincidence.

Nevertheless, one knows with a reasonable degree of certainty, Luciani went into the first conclave with a close margin of victory. One also knows, with a reasonable degree of certainty, Karol Wojtyla, the Opus Dei candidate, had the rest of them. That is, if one works backwards from the results which are known facts and not forward from the predictions which were pipedreams. For no one, other than the conclave cardinals, knows what goes on in a conclave.

---

[1] pg 156 *Le Soir Brussels* 10 Aug 78
[2] pg 156 *London Times* 9 Aug 78 Yu Pin's death is covered on the Internet
[3] pg 157 *L'Osservatore* 22 Aug 78 *La Stampa* 25 Aug 78

Chapter 11

His Papacy

Just a month after Luciani sent his widely published letter to Louise Brown that had embarrassed many cardinals who had condemned the child, Pope Paul VI died. On August 12, 1978, the *London Times* published a list of the leading candidates for the succession. In order of their promise were listed cardinals Benelli, Siri, Hume, Pignedoli, Baggio, Poletti, Lorscheider, Koenig and Cordeiro. Conspicuously absent was the name of Albino Luciani.

Astonishingly, when the white smoke rose from the Sistine Chapel on August 26, 1978, it bore his name. He was elected with ninety of the one hundred eleven votes cast. On the fourth ballot the count was seventy-five for Luciani; a marginal vote, exactly the two-thirds majority plus one vote required to elect him. [1]

On the recount, in order to make the election unanimous, fifteen others went his way. Only the twenty-one members of Opus Dei and the *Vatican Curia* remained cast against him. In their loss, they were determined to send a message in their dissenting votes. He would be the only pope in the thousand-year history in which cardinals have elected popes, who failed to have carried a unanimous vote on the recount. So bitter was the hatred of those opposition cardinals, some of whom would share the Vatican with him, they could not accept the fact that Christ had spoken.

In the eyes of the public and the press, Luciani's election was a remarkable occurrence, in that he was an outspoken liberal and it was thought less than a third of the voting cardinals had liberal tendencies. Most puzzling of all was that he had so recently defied a papal decree; an action that most thought would cost him his *red hat* let alone his chance at the top. He had publicly upstaged and embarrassed many of the voting cardinals who had condemned the world's first artificially conceived child.

This was a logical conclusion on the part of the public as all of Paul's public actions had moved only about a third of the *College* to the *left*. Yet, the cardinals themselves knew otherwise.

Ten years before, in 1968, Paul visited Luciani in Vittorio Veneto. On his return to Rome, Paul issued the doctrine *Humanae Vitae;* the Church's ban on contraception under any circumstances.

The very next week, Luciani went to the press with his famous letter demanding contraception be permitted in those cases where it was justified. In his letter, Luciani cited the massive poverty and starvation a ban on contraception would bring about in third world countries.

About a third of the voting cardinals were conservatives who were either from these countries or were sympathetic with them. In issuing the proclamation and instructing Luciani to follow with his public rebuttal, Paul ingeniously created the platform of Luciani's candidacy which ten years later would win for him the papacy; although conservatives in every other way, they felt so strongly of the havoc the doctrine was causing their congregations, they would vote for the man most likely to repeal it, Luciani.

John Paul

In accepting his pontificate from the *College of Cardinals,* Luciani took the name John Paul. History has recorded he named himself in honor of his patrons - John XXIII who had groomed him and made him a bishop and Paul VI who had made him a cardinal and paved his way to the papacy. Yet, one can surmise that he also had in mind his beloved father, Giovanni Paulo Luciani - John Paul Luciani - who had sheared his wool and groomed his mane. His father had been born Giovanni and had taken the name Paulo in Confirmation. In any event, he certainly covered the bases.

Accepting his papacy, he made a simple statement, *"Moses may never forgive you for what you have done today."*[2] This was his actual statement as reported in the local press. Later, foreign newspapers and some authors, who wrote about John Paul, took literary license and changed the word 'Moses' to 'Christ' - taking the name Moses to be synonymous with the name Christ.

If one studies his life, Luciani very obviously did not mean Christ. John Paul meant exactly what he said in this, the first comment of his brief papacy, *"Moses may never forgive you for what you have done today."* It was meant as a forewarning to the *Curia,* those twenty or so cardinals who would share the Vatican with him, their days were numbered; he intended to bring an end to the bigotry Moses had led the world into these past four thousand years. It hinted that the new Pope might make his move in the not too distant future; he would lead the Church and the western world away from Moses and closer to Christ.

A simple man

Avoiding the pomp and pageantry traditionally surrounding the installation of a pontiff, he took his office in a small private setting witnessed by the minimum number of Church prelates required and by his family and close friends including my friend Jack who had served him so faithfully at Vittorio Veneto.

Outside, a huge crowd, which had filled St. Peter's Square, kept its eyes watchfully on the balcony anxiously awaiting his first blessing as pontiff. But no one appeared; just a rustling, a kind of a murmuring of the crowd that started at its edges and eventually filled the square. Luciani had chosen not to display himself from the royal balcony as all the others had done before him. Rather, he had chosen to walk among them.

In taking his place as the leader of the Roman Catholic World with far less ceremony than that which had accompanied his installation as a common bishop twenty years earlier, he had begun to demolish the majestic image of the papacy. He refused to be crowned with one of the gold and jewel encrusted papal tiaras, which had been the focal point of previous coronations. [3]

In fact, there was no coronation at all. Instead he allowed a simple pastoral stole, the symbol of a common priest, to be placed upon his shoulders to mark his assumption of his responsibilities. He did not intend to rule from the throne. His peers, the cardinals, the crown princes of the Church, felt much of their own regency endangered. Whereas the rank-and-file and the hierarchy of the Church saw in the papal tiara a symbol of royalty, Luciani saw something much different. He saw in it the right to a good and healthy life for a thousand children who would otherwise starve to death, and that's exactly what he intended to do with it.

Actually, Luciani had never been a man of formalities. As a bishop, he had refused to be addressed as *Your Excellency*, the title reserved for bishops. As a cardinal he had refused to be addressed as *Your Eminence,* the title reserved for cardinals. And as a pope, he refused to be addressed as *Your Holiness*. He asked that everyone, from heads of state to little children, address him by the nickname he had acquired as a child, Piccolo.

He had been the first bishop installed by John XXIII and at the conclusion of the ceremony in St. Peter's his sister Antonia, who headed up the congratulatory reception line, approached him and motioned to bow to kiss his newly acquired bishop's ring. So horrified was he that this woman who had been such great support to him during his growing years would bow to him, that rather than extending his hand, he grasped her in his arms and held her in an embrace that spanned several minutes. The entire congregation was moved to tears. A group of Vatican cardinals, appalled at what they had just witnessed stood off to one side, frozen in hostile stares. Finally, on releasing her, this newest prince of the Church, realizing that he had broken Church protocol, turned toward the Pope and rather than apologizing for his action told him, *"A prince must never forget that his sister is the princess!"* He never donned the Fisherman's Ring, the symbol of the majesty of the papacy; nor was he ever known to extend his hand for the ceremonial kiss. He would have no man or woman bow to him. Rather he would embrace his visitor, not in a ceremonial way, but in a real way.

There is also the record of his participation at mass. In his twenty years as a bishop, dressed in a common hassock, he said mass faithfully once a week; every Wednesday morning.

He would have recent graduates of seminaries say mass on Sundays in the great Basilica San Marco. Not once, no matter how formal the occasion, was he known to have taken his seat on the bishop's throne. Rather, dressed in a simple smock, he would serve as the young priest's sacristan. He would kneel before the priest and receive his blessing and communion; a practice which often cause the youth to tremble. He would tell him, *"Do not be afraid. In the eyes of Christ, all of his creatures are equal."*

He carried this practice into the Vatican. He would have young priests of the Gregorian Seminary in the Vatican Gardens say mass in his private chapel. Again, he would serve as their sacristan. He would say mass each Wednesday in various surrounding churches to be closer to his congregation. Even in formal audiences, he dressed in a simple smock.

Photos have survived that show him dressed in elegant attire. Yet, if one looks closely, one will find that most of these have been doctored; a part of Rome's strategy to destroy his image.

He states his case

His positions concerning bigotry, *Planned Parenthood*, abortion, contraception, unwed mothers, out-of-wedlock children, remarriage, homosexuality and the poor were quite apparent in his formal acceptance speech in the Sistine Chapel on August 27, 1978, *"A particular greeting to all who are now suffering throughout the world; to the sick, to prisoners, to exiles, to the persecuted, and particularly, to those upon whom restraints are unfairly placed by doctrine in their everyday lives.*

"We must encourage all young adults in our congregation, no matter how scorned, to seek out and enter into long term loving relationships in order to provide for the economic and loving support of unparented children. We must take upon our shoulders responsibility to control the world's population, to bring an end to disease, poverty and starvation.

"Let our differences mold into one and together we shall rise to bring the world to a condition of greater justice. We call upon all of you, from the humblest who are the underpinnings of nations, to heads of state. We encourage you to build an efficacious and responsible structure for a new order, this one more just and honest... together we will muster the strength to lift those restraints that have been unfairly placed upon the everyday lives of so many innocent people by doctrine... for God-given human life is infinitely more precious than is man-made doctrine."[4]

He sizes up the Vatican treasures

In his first executive action, he ordered a complete review of the Church's finances, including a tally of all of its worldwide liquid assets. In fact John Paul, who had a background in finance, participated in the internal audit of the Vatican Bank himself. After all, it was he, who twenty years before, had inherited the bankrupt mountain diocese of Vittorio Veneto and had brought it to solvency and eventually to prosperity.

It is known he spent much of the month of his pontificate in the bank and he also walked around many of the other offices in Vatican City to see who was doing what. It was his personal involvement in the audit that led many authors, who subsequently wrote books that suggested foul play in his death, to point to bank officials, three of which, including the president of the Vatican's correspondent bank, eventually became victims of unexplained suicide or murder. Another disappeared and was never found.

About this same time, Luciani invited a number of art dealers to Rome for the purpose of obtaining appraisals of some of the art treasures of the Vatican Museum and the Sistine Chapel. It is also known that during his short reign he permitted a large real estate firm from Milan to survey the sprawling papal estate at the Castel Gandolfo on the outskirts of Rome.

Papal residence at the Castel Gandolfo resort

The papal summer residence boasts one of the finest wine cellars in the culinary world and is staffed with a world class culinary kitchen and chefs. The surrounding gardens are perhaps the most beautiful in the Western Hemisphere.

The Castel Gandolfo houses not only the summer residence but includes four other majestic palaces and gardens that are enjoyed by European cardinals and bishops when vacationing there. Actually it was a luxury resort city in itself. It was quite obvious from the beginning John Paul intended to make Mother Church heed Christ's most prolific testimony, *"If thou be perfect, sell all that thou hast and give to the poor."*

Luciani could not bring himself to accept the immense wealth of the Church; that he as its leader would live in luxury surrounded by priceless art and architecture and jewels and gold and feather pillows, while children in Africa and other parts of the world were starving to death. It anguished him much that it was the Church's position on birth control that was the reason why they were starving to death.

The equality of women

His position concerning the equality of women in the Church is made clear in the description of Luciani's last supper in the sequel to this book *Murder in the Vatican*.

"There were ten at dinner in all: the Pope, Villot, Casaroli, Caprio, the Pope's secretaries Magee and Lorenzi, and the four nuns who cared for the Papal Apartment... The Pope arrived first and pointed to a chair as each of the others entered the room...
"When they were all seated, Sister Maria Elena sat at the head of the table. Who's she? She was the nun one would often pass scrubbing the floors in the palace building. They did not understand this man who would have a scullery maid at the head of his table. Yet, it was not necessary they understand him. All that was necessary was that he understood himself.
"With everyone seated, the Pope gestured to Sister Maria Elena to begin grace. In the fifteen years of Paul VI' reign and in the five years of John XXIII' reign and in the nineteen years of Pius XII' reign and in the fifteen years of Pius XI' reign and in the eight years of Pius X' reign and in the fifteen years of Leo VIII' reign only two people had ever said grace in this room, the Pope and, in his absence, the Secretary of State. What kind of Pope was this who would have a mere woman lead the prayer?..."

In order to calm the fears of those conservative cardinals who shared the Vatican with him, who hated what he stood for so much, they had refused to vote for him, he did something that no other newly appointed pontiff had ever done before. He confirmed the appointments of all existing cardinals.

On the other hand he reduced in half the substantial bonus Vatican cardinals receive upon the election of a new pope. This seemed to be a forewarning to his eventually reducing the salaries of Vatican cardinals, which at that time was the equivalent of what is one hundred ten thousand pre-tax dollars today; spending money for the cardinals as all of their living expenses were paid for by the Church, most of them living in the lap of luxury.

This was something that Luciani's successor believed in. John Paul II raised Vatican cardinal salaries by eighteen percent. The action came so soon after his election it drew a comment from a *leftist* cardinal that reached all of Europe, *"It is almost as if it had been part of the deal."*[5]

Also, it was well known in Europe that upon becoming Patriarch of Venice, Luciani had reduced his living quarters to a small four room apartment in the rear of the fifth floor of the patriarch palace and had converted most of the remaining part of the palace to quarters for unwed mothers. There was considerable apprehension among field cardinals that they might end up sharing their sprawling mansions and palaces with the homeless and the poor. Worst of all, there was an underlying dread among all cardinals and bishops of the Church that they might be soon taking a vow of poverty. It might have occurred to some of them that they might soon be giving up their porterhouse steak.

The bishops of the Church were very aware of his threats to liquidate the Vatican treasures and impose a vow of poverty on all clergy and his intent to bring about equality of those who followed *Planned Parenthood,* women, the remarried, homosexuals and others whose everyday lives are scorned by doctrine.

A modern day Martin Luther

There was from the very start, the likelihood that Luciani would appoint a new cabinet; replace those who held the most powerful positions in the Roman Catholic Church. Shortly after his election, Bishop Casaroli, the Vatican Foreign Minister, was reported by an Italian tabloid as having said of Luciani, *"It is as if Martin Luther has come back from the dead to take his revenge on all of us. First to take from us our bonuses and soon our salaries* and then our rank and possibly even our palaces and then our porterhouse steak; perhaps it will end in our excommunication."[6]

There was some truth to this, for Luciani had studied Luther's life and often spoke of him. When he became bishop and overseer of the seminary at Vittorio Veneto he added a subject, *The Reformation*, to the curriculum and he, himself, taught it.

Most people think Martin Luther's central purpose in his *Thesis 95* objected to the sale of indulgences and absolutions by the Church. Actually, Martin Luther's basic premise was much broader than this. As Luciani pointed out in one of his texts *"Martin Luther believed that man's relationship with God was a private and direct one and that a middleman was not necessary to one's salvation, particularly, a middleman who was driven by hatred and greed. To put it pointedly, his goal was to do away with the idea of a church once and for all; despite the fact that today's*

Lutheran Church claims to have grown out of his teachings.

"Although he succeeded to some extent in accomplishing these things, Luther did not foresee that ruthless and greedy men – including many of today's evangelists - would seize upon the opportunity and begin to form their own churches in order to spread their hatred of those who appear to be different. They would capitalize upon the ignorance and weakness of the minds of men to spread their bigotry - the slavery of blacks - the subordination of women - the persecution and annihilation of homosexuals, Muslims, Jews and other ethnic peoples. Thus, we have a history in which Christianity has been a cesspool of hatred of one's fellowman which continues to go on today."

His forum

John Paul's public comments were largely limited to his audiences and we have already covered the lion's share of these in Chapter 1. When he said *". . . God is more our Mother than She is our Father,"* the press inquired whether this was a change in doctrine, that possibly a fourth person was being added to the *Holy Trinity*? Little did it know that Luciani intended to bring about a much more radical change concerning the *Holy Trinity*.

On one occasion, during a private audience, he took a microphone out of a cardinal's hand and handed it to a little boy as if to suggest that which the youngster had to say was more important than what his prince of the Church had to say. He had a particular knack for explaining complex issues in a simple and understanding way. On that occasion, he was asked by a rather frustrated eight-year-old boy, *"What is the difference between the left and the right? Why are they always fighting with each other? Is it because one believes in God and the other doesn't?"*

"No," the Pope answered, *"They both believe in God. It is just that those on the right want to guide their lives according to scripture, what someone is said to have said to someone else thousands of years ago."*

"Oh," said the child, puzzled by the Pope's reply. He then asked *"Then what do those on the left use to guide their lives?"*

"This," answered the Pope, *"This, their conscience,"* the Pope pointed to his temple.[7]

When interviewing an old woman who had described him as the greatest pope, a reporter asked her the reason for her appraisal. She replied, *"Because I understand what he says. I have known the rules all my life, but never before have I known why."*

When John Paul shook hands with the Communist mayor of Rome, the Vatican cardinals shot vicious glances and when he hugged him they shrugged in despair. On one occasion, John Paul was reported in the tabloids as being seen in St. Peter's Square, walking among the people wearing shorts and sandals.

Regardless of the validity of this reporting, he asked, he listened, he learned. He talked, he told, he taught. He grinned, he smiled, he laughed. Above all, he hugged. And best of all, they learned to hug him back.

He made few friends among those of rank in churches and nations, yet, he quickly won the friendship of the common man.

---

[1] pg 158 *Associated Press* 13 Sep 78
[2] pg 159 *L'Osservatore Romano* 27 Aug 78
[3] pg 160 One of several Papal tiaras
[4] pg 162 *L'Osservatore Romano* 28 Aug 78
[5] pg 165 *London Times* 28 Aug 78
[6] pg 166 *London Times* 21 Oct 78
[7] pg 167 *L'Osservatore Romano* 5 Sep 78

NOTE: See Chapter 1 for papal audience excerpts

Chapter 12

The Strange Visit of Metropolitan Nikodim

There is the strange visit of Metropolitan Nikodim to the Vatican in September of 1978. Nikodim was the Archbishop of Leningrad and youthful leader of the Russian Orthodox Church. One must keep in mind that Luciani was a firm believer in Christ. He believed much of the philosophy of *Communism*, Christ's philosophy, particularly as it pertained to a redistribution of wealth society, could be indoctrinated into modern civilization; something that had not happened in Russia.

Christ's enemy was *Fascism*, not *Communism*. Nikodim had met both Luciani and Cardinal Suenens quite by accident while traveling on a train from Rome to Florence in 1976. He became an important part of Luciani's plan to bring the philosophies of Christ into modern civilization. A job that Christ, Himself, had failed to accomplish in His ministry, as His alleged followers chose to embrace His great enemies - greed and bigotry.

Almost immediately, Nikodim became a powerful voice calling for the redemption of the American Christian,

"Here in my part of the world I must take my orders from the Kremlin bosses, certainly, if not at least I am at no necessity to pretend that they and I pursue the same end. For me, it is that which is told me by both my patron saint of state - Marx; and my patron saint of faith - Christ. So there is no inner conflict there for me.

"There is no originality in the works of Marx. His sole ambition was to make the voice of Christ a reality in a modern world. Yet, on the other side of the world, the American part of the world, the situation is much worse. There Christians foolishly accept the Devil's offer [1] *of the Kingdom of Heaven on the sole commitment that they fall down on their knees and worship him. Having once bred hatred of Jews and blacks and now having had that taken from them, they breed hatred of homosexuals, Muslims, atheists and others. They spit in the face of Christ and dare call themselves Christians.*

"America claims to be a free nation, yet, it deprives its citizens of the most sacred right of all - the right of privacy – the right to be left alone. It invades the bedrooms of homosexuals and imprisons them in some cases for the rest of their lives for doing nothing more than loving another human being. It invades the private homes of honest citizens and overturns card tables and trots entire families off to prison, while its holier-than-thou Christian brothers, who impose its gambling laws, openly play BINGO for profit in church halls. It invades the private homes of mothers and fathers and takes them away from their children for long terms for doing so little as being caught with a single marijuana cigarette which couldn't hurt a fly. Yet, its Christian leaders finance their elections with contributions from cigarette companies which kill tens of millions.

"The ink was not yet dry on the Civil Rights Bill when rightwing anti-black fanatics started their drug war which in the end will expand the proliferation of drugs and guns beyond anything civilization has known. The United States is now on a path toward imprisoning millions of innocent people that any other free nation in the world would treat in hospitals. America believes that the solution to social problems is to build churches and prisons rather than building schools and hospitals. The reason it thinks this way is to keep subordinate peoples down.

"But, perhaps, what demonstrates most of all, that America is not the land that it pretends to be, is its open borders to its rich neighbor to the north versus its closed borders to its poor neighbor to the south. It dares calls itself a Christian nation, as if to say that Christ would slam the door on these poor and starving children. Even today, ignoring the fact that medical science has determined who one falls in love with is God's will, the deranged minds of American evangelistic parents cause thousands of homosexual teenagers to take their own lives.

"Americans also falsely claim that other most sacred freedom - freedom of religion. There, freedom of religion means the right to force one's beliefs on others. In Russia we look at it quite differently. Here, freedom of religion is the right to believe or not to believe. Yes, it has its restrictions as we are free to practice our religions within the privacy of our homes and churches. Both Christian church bells and Muslims prayers on loud speakers that

broadcast religion are banned. Also, preachers are often imprisoned for preaching hatred of any kind of people, whether it be a matter of race, creed, gender or homosexuality.

"In America, on the other hand, church bells ring out at will carrying the message of Christianity; yet, Muslim Prayers over loudspeakers in Muslim neighborhoods are banned and violators are hauled off to prison.

"American tax laws, which grant tax exemption to wealthy churches, in themselves violate the basic concept of freedom of religion, as they impose unfair tax burdens on the non-believers. The non-believer pays in higher taxes for the believer to believe, and the believer uses his tax advantage to persecute the nonbeliever. Tax exemptions provide churches with tens of millions of dollars to finance radio and television time spreading their hatred of homosexuals and others who don't conform to their rules; depriving them of equal rights under the laws.

"Free? Free? Americans have not the slightest conception of what the word means. Although we here don't pretend to have it, we at least know what we are striving for - free, to be so free as to not allow the freedom of one to impose on the freedom of another.

"Although we too, to a far lesser extent, are guilty of some of these kinds of injustices, we being men and women of integrity do not claim to be a free state. American Christians are Pharisees - the Hypocrites that Jesus threw out of His temple. They disguise themselves in His name, yet cast aside His instruction." [2]

At the time, the laws concerning most of what Nikodim spoke of were more lenient in the Soviet Union than they were in the United States. Laws in some states, for example, permitted imprisonment of consenting adult homosexuals in private for sentences up to life. Although some homosexual and heterosexual acts were also illegal under Soviet statute, the laws limited imprisonment of offenders to a period of not more than three years, and the infraction had to be proved to be public and not private. In addition, in Russia, there is no record of a gay teenager having taken his or her life, although it may have happened. After all, there were no evangelistic preachers there to spread hatred of these innocent teens.

In 2003, sexual acts between consenting adults in private were decriminalized in the United States; a quarter century after Europe had led the way. The action to overturn its laws was made possible by heterosexuals who realized that many acts that they themselves were involved with in the bedroom were prohibited by state sodomy laws. Had President Clinton had his encounter with the intern Monica in the state of Alabama instead of the District of Columbia, he would have been guilty of a felony punishable for a term not to exceed twenty years. Luckily for him the District of Columbia had no sodomy law prohibiting oral sex.

The Supreme Court's decision overturning state sodomy laws which legalized sexual acts between consenting adults in private was a milestone event in the history of mankind. The sacred right of privacy had been a freedom in post-Christian Russia from the time of Lenin; it had been a freedom in the Mid East until the seventh century when Mohammed incorporated some of what Moses had to say concerning sex into his *Koran*; it had been a freedom in the pre-Christian Roman Empire; it had been a freedom in the Greek Empire; it had been a freedom in the Egyptian Empire dating back to 3100BC; it had been a freedom in China and India from the beginning of time; it had been a freedom in all primitive societies which survived unscathed into the twentieth century, including the African tribes to the south and the Eskimos to the north; it had been a freedom in the days of the Cro-Magnon Man.

Of all the scriptures of the world, only the *God of Moses* declares sex to be sinful and evil - the most prolific and powerful message of the Old Testament. Of all the men that ever lived only Moses and his mindless followers believed the way God makes babies is evil - the work of the Devil. All other scriptures of the world including those of the Hindu God Brahma, Buddha, Tao and the various gods of the Egyptians, Greeks and Romans all believed that sex is good and beautiful - a gift from God.

Today, no issue divides church from state more decisively. The Bible continues to hold that all sex is sinful and evil - the work of Satan; while the state has made the rounds and has now returned to the days of Pharaoh and Buddha and Tao - back to the days of the caveman - when sex was believed to be a gift from God.

Two cups of coffee and a few almond cookies

On the morning of September 5, 1978, forty-seven year old Metropolitan Nikodim was led by Mother Vincenza into the Pope's private quarters. This was quite unusual as John Paul normally met with visiting dignitaries in his public office. She carried with her a tray of two cups of coffee and a few almond cookies. Later the Pope told the press, *"After Sister Vincenza had left the room and closed the door, the Metropolitan reached for one of the cups and lifting it to his lips found it to be too hot and returned it to the table. He then fell forward onto the table dead."* [3]

It may be, the Pope told the press Nikodim had not tasted the coffee in order to quell rumors of poisoning. Yet, whether the Pope had told the press the truth or not is immaterial. We have yet another man of extraordinarily good health suddenly dropping dead of an alleged heart attack. Nikodim was a marathon runner and had just that past spring won an event. His body was returned to Leningrad the same day and because of the Pope's testimony the Russians never performed an autopsy.

Despite John Paul's statement to the press, rumors spread throughout Italy the poisoned cup had been intended for the Pope. If the coffee had indeed been laced with cyanide or a similar lethal toxin and had it been steaming it would have had the same effect as a cyanide capsule, which vapors result in immediate paralysis and death. Yet, had Nikodim actually tasted the coffee, it could not have killed him so quickly as there is no poison when ingested that kills its victim instantly. Even potassium cyanide results in a series of convulsions before death. Unfortunately, the cups were washed before rumors surfaced. Nevertheless, it is a strange coincidence the Pope, himself, would die just a few days later of identical symptoms; as if he had bitten down on a cyanide capsule.

The Metropolitan had been summoned to Rome to head up a venture that would change the economic psychology of the western world, which was at the time and still is today driven by Christ's number one enemy - greed. His job would be to change it into a world of helping others. The Metropolitan would work side-by-side with John Paul to achieve their mutual goal - to bring western civilization away from the greed and bigotry of the *God of Moses* and closer to Christ.

The arrangement was to be twofold. John Paul would set the example by liquidating the Vatican treasures, perhaps going so far as to turn Vatican City over to real estate developers for tens of billions of dollars and establishing modest quarters in the countryside. Nikodim would follow with his part of the plan - to change the psychology of the world by converting modern day Christians from the hypocrites they are into more compassionate human beings; Christ-like human beings.

That each time one would come to the fork in the road, one would no longer ask oneself, *"Now, what is in this for me?"* but rather one would ask oneself, *"Now, what is in this for others?"* This would be a horror worse than death for today's Christians. A monstrous job, yet one that Nikodim was both physically and mentally equipped to carry out. That is, before he dropped dead.

As already mentioned, it was the sudden death of Nikodim and the close alliance John Paul had with Communists that caused some authors who later wrote about John Paul's death to implicate American and British intelligence agencies in foul play. Certainly, a capitalistic society thrives on greed and it would not be in the best interests of these countries to have the most influential man in the western world preaching a redistribution of wealth society.

There is some sense to this. After all, when one considers the fundamental ideology of *Communism*, the equality of all of God's children and a redistribution of wealth society, the world has never known a more steadfast communist than Pope John Paul I. Yet, there are at least two holes in this theory. It is doubtful that these agencies could have acted this quickly as Luciani was only a month into his papacy; one that he was not expected to win.

What nails the lid shut on this theory is that Jimmy Carter was president at the time and he would have welcomed a more compassionate society. Had it happened at a time when America had a president who himself is driven by greed, whose every move is to make the rich richer and the poor poorer, any court in the world would have indicted him for the murder of John Paul I.

---

[1] pg 168 *"Devil's offer"* Nikodim is referring to the *God of Moses* of the Old Testament
[2] pg 170 *Leningradkava Veteran* 17 Oct 76
[3] pg 172 *L'Osservatore Romano* 6 Sep 78

Chapter 13

Dark Night - his death

Thirty-three days after his installation, he died mysteriously in the Vatican. He was found dead in his bed early on the morning of September 29, 1978. Strangely, although death was determined to have occurred before midnight, he was in his daytime clothes and in a sitting-up position. The bed lamp was still on and the windows were wide open. Still upright in his hands were notes he had written when he had been Bishop of Vittorio Veneto.

At seven-thirty in the morning, the Vatican released an official statement, *"Pope John Paul died just before midnight, September 28, 1978, of a blood clot to the heart. He was discovered by his secretary John Magee about six-thirty this morning when he went to look for the Pope when he failed to show up for his morning chapel service. He was found to be in a sitting up position and wearing his daytime clothes and the lights were on. John Paul died while he was reading the 'Imitation of Christ' which book was still held upright in his hands. Father Magee, on realizing the Pope was dead, summoned Cardinal Villot who performed the last rites of the Church."*[1]

Papal bedroom (top right corner) looks down on St Peter's Square

Newspapers immediately questioned the release. Had the Pope's light been on all night, it would have been an unusual occurrence and would have been reported in the press the next day. Scores of tourists and reporters roam St. Peter's Square into the wee hours of the morning. Also, had police who were assigned to the square noticed the Pope's light on all night they would have reported it to the Swiss Guard who would have investigated it.

The release also stated that he was reading the *Imitation of Christ* when he died. However, when asked by the press, the embalmers, who took papers from his hands, not thinking it to be an important issue, told the reporters that in his hands were some notes written on the stationary of the Diocese of Vittorio Veneto. On the bed were spread a few old newspaper clippings.

The embalmers, the Signoracci brothers from a nearby school of medicine, told the press they were picked up by a Vatican car at a little after five-thirty in the morning which was an hour before the Vatican release announced the body was found. They also told the press they had been told by Swiss guards that it was a nun who had discovered the Pontiff early in the morning. In addition, it was their opinion the Pope had not been dead for more than an hour or two, as it was a rather chilly morning and the windows were wide open and the body was still warm.[2]

The official Vatican release, itself, had confirmed that the Pope had died in the early morning hours and not before midnight in that it had specifically stated Cardinal Villot had performed the last rites. No priest, let alone a cardinal, would perform the last rites over a cold corpse, as Church doctrine decrees the soul leaves the body when rigor mortis sets in.

The press was quick to follow up on the discrepancy as to who discovered the body and interviewed Sister Vincenza the nun the embalmers claimed had found the body. She told them the Pope normally rose between four and four-thirty and she routinely delivered coffee between four-thirty and quarter-to-five.

The testimony of the nun who found him

When she first knocked on the door there was no answer. She waited a minute or so and knocked again, this time a bit louder. It was obvious to her the Pope was still in the bathroom; perhaps he may have risen a bit late. She opened the door and entered the room, intending to leave the tray on his nightstand.

It was then she saw the bed lamp was on and he was in his daytime clothes. He was sitting up reading some papers he held in his hands. She greeted him with *"Good morning."* He did not respond. He held to what she described as a mime position as if he was too deeply involved in what he was reading. It was not unusual for him once dressed for the day to be sitting up in his bed reading when she delivered breakfast.

Vincenza had come to know this man as a jovial one, always smiling, often laughing and sometimes joking. So at first she thought this was some kind of a joke. Particularly, since he had a kind of a leering grin on his face.

Approaching the bed she said, *"Please don't joke with me in this way, Albino."* It was when she placed the tray on the nightstand she realized something was wrong. She immediately went out of his room and down the hallway and fetched John Magee who came to the room and found the Pope was dead.[3]

Her testimony confirmed that of the embalmers. The Pope was, in fact, holding stationary of some sort and not a book. It also pointed to a discrepancy in time, as the Vatican release had stated the Pope had been found at six-thirty in the morning. John Magee had been the secretary of his predecessor Paul VI and had been serving in transition.

The mysterious notes held in his hands

These contradictions of the official release of the Vatican gave birth to a rumor that the notes held in the Pope's hands were listings of cardinals to be replaced; John Paul was planning a shakeup of the Church's hierarchy. Although it certainly would not make very much sense for him to be using outdated stationary for such a purpose, the rumor spread like wildfire. It would make more sense, at least to this author, John Paul was in fact reviewing a draft of one of the many controversial letters he had written while he was Bishop of Vittorio Veneto. It would seem most likely to be a draft of his famous letter to Paul VI requesting revisions be made to the doctrine of *Humanae Vitae*, the Church's policy prohibiting the use of contraceptives; one it was expected he would repeal. His letter of 1968 to Paul VI recommended "*. . . Accommodations for birth control must be made within the confines of the Church.*" It would seem since it had been a public letter, in his proclamation, he would say nothing that might contradict what he had said ten years before.

This would also explain the newspaper clippings that were found on his bed, as his letter had been widely published in Europe. Then, again, it could have been a draft of any public letter he had written during his time as Bishop of Vittorio Veneto on a wide range of issues which included divorce and remarriage, equality of women in the Church or even homosexuality; all issues which horrified Opus Dei and the *Vatican Curia*, that cluster of Church prelates which shared the Vatican with him, a few of which shared the Papal Palace itself with him.

Perhaps, most threatening of all, he might weave the most prolific message of his ministry, *"We think of sex as the greatest of sins, whereas, in itself, it is nothing more than human nature and not a sin at all,"* into *Canon Law*. Then there was expectation that he would restructure the *Holy Trinity*. Something he had hinted at the day before he died, *"God is more our Mother than She is Our Father."* Nevertheless, to this day, the Vatican has never released the nature of the notes he held in his hands.

The early morning embalming

The official Vatican release had specifically stated the Pope was discovered at six-thirty in the morning. By six-thirty, the body had already been embalmed. Cardinal Villot, recalling the foul odor that the body of Paul VI had emitted after his death just a month earlier, ordered an immediate embalming of John Paul's remains in order to prevent a recurrence of what had happened to Paul, which in his case had delayed the body's viewing in St. Peter's Basilica by a day. This explains why the embalmers were brought to the Vatican so early in the morning.

Yet, that the body had been embalmed so quickly gave rise to a rumor the Pope had been poisoned. It was commonly known this was a practice of the Mafia when slow arsenic poisoning was the instrument of murder; embalming the body shortly after death erased the obvious signs of poisoning. It was for this reason that it was illegal in Italy to embalm a body until twenty-four hours after death. Of course, Italian law did not apply within Vatican walls.

After John Paul II was elected, in order to quell the rumors, Lorenzi told the press the Pope had complained of chest pains at dinner. Instead, his testimony fired the rumors as there are many lethal toxins that can precipitate heart attacks in healthy people. Also, heart attack would mean someone dressed him up and sat him up in bed in the position he was found by Magee and Vincenza. Also, his testimony contradicted that of Sr. Vincenza who had been sitting next to the Pope. She had told the press the Pope was laughing and in good spirits at dinner.[4]

Sequestering the witnesses

When all this conflicting testimony was questioned by the press, the Vatican finally issued a verbal corrective statement to the press that in the confusion following the Pope's death some errors had been made in its original release.

Because the Vatican had been caught in what appeared to be a combination of lies, rumors of foul play continued to spread throughout Europe. The one rumor that seemed not to go away was that the notes held in the Pope's hands at the time of his death were lists of cardinals to be replaced; John Paul was planning a shakeup of the Church's hierarchy

Another question emerged. If the Pope retired at nine o'clock, how could he still be in is daytime clothes and be reading at midnight? When John Magee answered this question he told the reporters that he had forgotten to wake the Pope from his nap that day; the Pontiff had slept past six o'clock. This would explain why he could still have been reading at midnight, as he would have had trouble getting to sleep.[5]

This raised even more suspicions, as the Pope was never before known to take a nap. Again, Magee responded that the Pope had started taking naps about two weeks earlier; perhaps the strain of the workload was too much for him. This caused even more concern, as one of the things that evidences early stages of slow arsenic poisoning, as well as many other poisonings, is tiredness; one might take unusual daytime naps.

On the third day following the Pope's death, all those who had shared the Papal Apartment with the Pope were reassigned to unknown locations in Europe. This included the four nuns who cared for the apartment and the Pope's secretaries John Magee and Don Diego Lorenzi. This meant that Sister Vincenza who had served him for twelve years would be unable to attend his funeral. This enraged reporters, as it removed the primary witnesses to the circumstances surrounding the Pope's death from access.[6]

Every newspaper in Italy questioned the Vatican's action in this respect. If it had nothing to hide why exile the only known witnesses to the circumstances surrounding the Pope's death? In an attempt to bring an end to the rumors, the Vatican finally responded with a corrected release to the press,

"Pope John Paul did not have in mind to make changes in the Vatican hierarchy as evidenced by the following facts,

On August 27th he reconfirmed Cardinal Villot as Secretary of State

On August 28th he confirmed all existing cardinals for their current five year terms.

The intention of sudden sweeping change would be entirely out of character for him. In his previous assignment as Archbishop of Venice he had kept the status quo.

The Pope's secretary and the nuns who cared for the Papal Apartment were particularly saddened by the Pope's passing and were placed on sabbatical leave to help them get through this difficult period.

We wish to correct our statement that the Pope held the Imitation of Christ in his hands at the time of his death. This was a communications error. The Pope at the time of his death was reviewing some old notes that he had written when he had been Bishop of Vittorio Veneto. They had been written on diocese stationary. That he was able to retain them upright in his hands in the midst of a massive heart attack is due to the grace of God

We wish to correct our original release that it was the Pope's secretary John Magee who had discovered the Pope's body. It was the Pope's secretary who first realized that he was dead. The Pope was first discovered by the nun who delivered the Pope's breakfast at the usual time. When the Pope did not respond she summoned

John Magee as she sensed something was wrong. It is Canon Law that an autopsy cannot be performed on a pope's body.

It is immaterial whether a nun or his secretary found His Holiness. It is also immaterial when he was found dead. And it is also immaterial when he actually died. All that is material is that he was found dead. The Vatican. [7]

The letter simply raised more questions. Reconfirmation of Cardinal Villot was meaningless as the Secretary of State is normally appointed for life. Yes, he can choose to retire, in which case another cardinal can succeed him, but otherwise he will normally remain in his position until he dies. The reason for this is the appointment by a pope of a secretary of state is deemed to be an infallible act as if a pope dies the secretary of state becomes the interim pope with all of the ecclesiastical powers of a pope. A successor pope cannot remove an incumbent secretary without endangering the infallibility of his own office.

Also, confirmation of the cardinals as cardinals did not necessarily confirm them in their present positions of power. It was widely rumored that several of them would lose their rank. It was speculated a new foreign minister would be named and several heads of important councils and congregations would be replaced. Also, several cardinals who held key positions as apostolic delegates to major nations, including Great Britain and the Soviet Union, would be replaced.

One should keep in mind that these speculations were at the very best leaks to the press. They were not known facts. Just what were John Paul's intentions were known only to himself; that he died a month before he was to make his first proclamations they will remain forever a mystery. Yes, that they would have been revolutionary was quite certain. But just exactly what changes he would have made had he lived remain unknown

The claim that radical change was not characteristic of Luciani came under attack. Although it was true that he had made very few changes during his time as Patriarch of Venice, it was widely known when he had become Bishop of Vittorio Veneto, ten years before, just the opposite had been true. Within a month of his assignment he had replaced both the president and the dean of the great seminary there. Within a year he had replaced more than half of the school's faculty. Actually, it was his record at Vittorio Veneto together with his liberal reputation that had inspired the rumor of change in the Church's hierarchy in the first place.

Also, it made no sense that Magee and three of the nuns would have been so saddened by the Pope's death that they were placed on sabbatical as they had known him only a month.

Although it was *Canon Law* an autopsy can't be performed on a pope's body, it does allow for certain exceptions. Suspicion of murder is one of these; the reason why an autopsy was performed on Pius VIII in 1829.

The slipper socks, spectacles and the lock of hair

There were a few other questions that were innocently raised by the Luciani family itself. The Pope's half-sister sought to recover a pair of slipper socks that she had knitted for the Pope in honor of his elevation to the papacy. She knew, as a boy, he had often gone barefoot in the Italian Alps; she wanted to make sure he had something to keep his feet warm. They were white and had his coat of arms embroidered in gold on them.[8]

Then there were his spectacles. Although the Pope did not require them for reading; which explained the mystery as to why they were not found on his person, as he had obviously died while reading, he would have required them just to walk across the room, as he suffered from acute near-sightedness.[8]

Yet, I suppose, if one wants to add these to the stack of unanswered questions concerning John Paul's death; yes, these items have never been found. Then there was the fact that his sister Antonia, in response to her request for a lock of his hair, received a clump of jet black hair two weeks after his interment, which she claimed could not have been his, as his hair was graying.[9]

There was the unanswered question as to how the Pope could have remained in a sitting-up position with the notes still clutched in his hands if he had, in fact, suffered a heart attack? Only the most gullible of believers accepted the Vatican's contention *"It had been by the grace of God."* The Vatican decided to change its policy. It took a position of no longer responding to the press.

The bell cord and the intercom

There remained the claim he had retired at nine o'clock, being unable to sleep, he decided to review some notes he had written in Vittorio Veneto. In that he died before midnight still wearing his daytime clothes is a puzzling happenstance in itself. If one decides to read oneself to sleep, one will invariably first don one's pajamas or other bedclothes.

Yet, most questionable of all, was that something killed him so suddenly he was unable to reach for the bell cord, which hung a whisker away from his right shoulder, as this would have summoned in an instant the guard who sat at his desk just a few feet away at the entrance to the corridor that led to the Pope's inner chambers. Also, he was unable to press one of the service buttons on the intercom that was on his bed stand immediately to his left that would have brought to his side any of five other people who resided elsewhere in the Papal Palace that night.

The Vatican newspaper itself had reported an interesting coincidence. On the second morning preceding the Pope's death, the Vatican maintenance workers happened to test the bell cord, something that had not been done for years, as Pope Paul always used the intercom. The article said that the bell rang so loud people who were in the palace at the time, including the Pope himself, thought it to be a fire alarm and headed for the stairs. [10]

The missing will

There was another strange occurrence. John Paul's will could not be found. This was particularly confusing, in that, as a part of standard indoctrination, an incoming Pope is required to file his will with the Vatican. When it was determined the will had been misplaced in the Vatican offices, his lawyer in Venice was asked to send a copy to Rome. At first he complied. Then a few days later sent a message to the Vatican he had discovered Albino Luciani's will missing from his files.

In that the will could not be found, meant that the Luciani family had no right to any of his belongings other than what the Vatican might option to give them. To anyone with suspicions of foul play, it meant that the family could not obtain anything that might be construed as being DNA today. Needless to say, the missing will added more fuel to the rumors that were spreading across Europe. To this day, Luciani's will has never been found.[11]

As already explained, the lack of a will gave the Vatican authority to confiscate and destroy anything he may have said that may have been controversial in all of his private papers from the time he was a child through and including the papers he held upright in his hands on his deathbed.

To this day, the Vatican has never revealed the nature of the papers he held in his hands – which revelation would have brought and end to the rumor that they contained listings of cardinals to be replaced – unless of course, they did contain listings of cardinals to be replaced. What one does know is that they contained something the Vatican did not want the world to know – otherwise it certainly would have released them.

There was the unanswered question as to how the Pope could have remained in a sitting-up position with the notes still clutched in his hands if he had, in fact, suffered a heart attack? Only the most gullible of believers accepted the Vatican's contention *"It had been by the grace of God."* The Vatican decided to change its policy. It took a position of no longer responding to the press.

Nevertheless, no explanation of this strange combination of events, or lack of events, was ever offered by the Vatican, or for that matter, by any of those who wrote books about the event.

[1] pg 174 *IL Manifesto* 29 Sep 78
[2] pg 175 *La Repubblica* 30 Sep 78
[3] pg 176 *IL Messaggero* 30 Sep 78
4 pg 178 *La Repubblica* 1 Oct 78
5 pg 178 *Il Tempo* 1 Oct 78
6 pg 179 *IL Manifesto* 3 Oct 78
7 pg 180 *IL Messaggero* 11 Oct 78
8 pg 181 *IL Manifesto* 13 Oct 78
9 pg 181 *La Repubblica* 15 Oct 78
10 pg 182 *L'Osservatore Romano* 27 Sep 78
11 pg 183 *Secolo d'Italia* 20 Oct 78

Above are the original releases in the press. They can also be found in world newspapers with some editing out by the Vatican of controversial things he said and did. It is not the purpose of this book to investigate the mysterious circumstances of the death of the 33-Day Pope and those around him. Yet, it is necessary to mention these things as they were an integral part of his life.

Chapter 14

The Deception

Shortly after his interment in the crypt beneath St. Peter's, there began a structured program by the Vatican intended to create two misconceptions, 1) he had been all his life a stanch conservative, 2) he was on the brink of death when elected to the papacy.

The conservative Luciani

Concerning the first of these, there began a massive effort to destroy the liberal identity of the 33-day Pope. Teams of Vatican workers showed up at Belluno, Vittorio Veneto and Venice scouring through his writings, sermons and public releases destroying anything controversial he might have said or done on a wide range of issues. In some cases, original documents were destroyed and replaced with forgeries to substantiate he took positions the exact opposite of what he actually stood for. There is not a single record in the Vatican today of dozens of his letters challenging the Holy See on humane issues before, during and after the World War. Letters one knows he did write as they gained notoriety in the press.

Among them is his letter to Paul VI of 1968 challenging *Humanae Vitae*, the Church's policy banning contraception. Despite the fact that excerpts of the letter were published in the world's newspapers, the Vatican denies that such a document ever existed. All traces of Luciani's communications as a priest, a bishop and a cardinal with the Vatican has vanished; as if to say a man who rose to the papacy never dealt with the Vatican. [1]

In its deception, the Vatican republished just about everything he said publicly or privately during his papacy including his audiences. In one case, the Vatican published a statement a month after his death that Luciani had warned against the dangers of artificial insemination in an address to cardinals in the Sistine Chapel a week after he sent his congratulatory message to the parents to Louise Brown. [2] Milan newspapers reported Luciani as being in Milan on the day he supposedly spoke in Rome. Even

Luciani's own books have not survived untouched. If one takes the time to examine his best seller *Illustrissimi* as published before vs. after his death, one will find subtle changes leaning him toward the *right*.

In the following years, came a flood of biographical briefs from Rome which purpose was to create the image of a man who had spent his lifetime on his knees and ignored the issues of his day. The intent was to change his image from the progressive he had been, to the conservative the Vatican wanted him to be remembered as; the purpose, to remove the ecclesiastical motive for murder.

His medical condition

Concerning his physical condition, there emerged a flood of rumors designed to establish Luciani as having been a man of extremely poor health to substantiate the Vatican's claim he had died of a heart attack. While the rumors of foul play in the Pope's death were still spreading across Europe, there came a strong rumor Luciani had suffered a half-dozen heart attacks during the last year of his life. In aspiring to be pope, so the story went, he had kept his deteriorating medical condition to himself because it would have affected his candidacy. There was some credibility to such a rumor, except for the known facts.

Here again, the reporters followed up on the alleged heart attacks and have left the record behind them in the microfilm. They found that the Luciani family had no knowledge of the alleged attacks. His own personal physician in Venice had never known of them. The press interviewed every cardiologist in the entire Venice metropolitan area and none could be found who had treated the cardinal. There was no record of him in any of the area hospitals. It would have been quite a trick to have suffered six successive heart attacks in a very short time and not have required the services of a cardiologist, much less a hospital. His secretary Lorenzi and the housekeeper who had shared his private apartment in Venice with him had no knowledge of the presumed attacks. There was the fact that his medical exam of six months before had showed him to be a man of extraordinarily good health. Finally, there was the fact that for five hundred and twenty six consecutive Wednesdays and for five hundred and twenty seven consecutive Sundays, at precisely seven o'clock in the morning as reads the face of the clock; Albino

Luciani had said mass for his congregation in Venice.

In addition, Rome's biographical briefs depicted a man who had suffered from severe respiratory illnesses all his life even going so far as to claim as a child he had been confined to a sanatorium for a year with tuberculosis. This *fact* has found its way into a number of books since, including some of those that cited foul play in the Pope's death. So cleverly does the Vatican drop its seeds of rumor, in some cases even those books attacking the Vatican's claim he died of a heart attack, innocently restate this rumor as being fact.

There is no record in Belluno today that a boy named Albino Luciani was ever confined in its sanatorium. Yet, there is nothing that proves he may not have been confined there. Nevertheless, medical science tells us in the early twentieth century tuberculosis, as it is in undeveloped countries today, was a killer. At that time, even with care in the most advanced sanatoriums, the survival rate was less than twenty percent. The chance of survival was infinitely less in malnourished areas like in which Albino had grown up. The Vatican's biographical briefs also claim Luciani was confined for most of 1947 in the hospital in Belluno with severe viral pneumonia.

In my visit to Belluno in 2004, the hospital had no record of either his rumored yearlong confinement to its sanatorium as a child or of this more recent confinement for viral pneumonia. It would be quite a trick for one to have suffered pneumonia for upwards of a year and survive. On the other hand, it did have a record of his hospital stays in 1926 for a tonsillectomy and in 1962 for a broken nose suffered in a soccer game and a gallstone operation. Strangely, his biographical briefs mention none of these.

Nevertheless, there is no chance he suffered from extreme respiratory problems all his life. We know this to be a matter-of-fact because he spent the last thirty years of his life as an accomplished mountain climber. When he struck his coat-of-arms in 1958 when he became a bishop at the age of forty-six, he included thereon the six Dolomite Mountain peaks for which he held the speed record. In the years to follow, he would add a number of the tallest and most difficult peaks of the Italian Alps to his achievements. Mountain climbing, as one knows, requires immense physical strength. The fundamental physical requirement of an accomplished climber is that one possesses a powerful respiratory system. The Vatican should have thought of this when it started its rumors that Luciani struggled throughout his life with death threatening respiratory problems; so much so it claims that he spent most of his adult life in and out of sanatoriums. Also, respiratory failure could have never killed him so suddenly.

Several field clergy scattered around the globe innocently helped the Vatican in its deception by writing biographical briefs in local languages that depicted a man who was on the verge of death most of his life. One of these was published in the United States by a Carmelite nun, *"When he was born the midwife who attended baptized him immediately because he was so frail, she feared he would die."* And later in the book, *"Within a year of his ordination, his health, never good, broke down and he was taken to the sanatorium, a very ill young man. The antibiotics, which could have helped him, had not yet been discovered, and Albino expected to die soon. He recovered but several times later he again spent many months in and out of sanatoriums."* On an ongoing basis for the remainder of the work, this innocent nun speaks of Luciani as being in such fragile condition as if he might not it make it to the end of her book. [3]

The grace of God

At the time of his death, John Paul had no known medical condition other than some unexplained swelling of his feet, which his doctor had diagnosed the week before his death as the onset of arthritis. Swelling of the extremities can be caused by an endless array of physical problems. Also, for those of you who have no substantial experience in reading mystery novels, swelling of the extremities - particularly of the ankles - is an early symptom of slow arsenic poisoning. Then there were the uncharacteristic naps, also symptomatic of slow poisoning.

His physical examination of six months before his death which was released to the press at the request of his family, had detected nothing other than a man of extraordinarily good health.

The position he was found, *"sitting up in his daytime clothes reading some papers he held upright in his hands,"* was the testimony of both the original and corrected Vatican releases, Mother Vincenza and Father Magee who found him dead, the embalmers and all of the others who were brought to his room.

Yet, the notice he had died a natural death was accepted by most rank and file Catholics. This demonstrates the vast power of the propaganda machine of the Vatican. In the face of overwhelming evidence to the contrary, it has convinced not only its own congregation, but most of the world, John Paul died of a heart attack. It has convinced the world John Paul was unable to reach for the bell cord a whisker to his right. It has convinced the world he was able to retain the exact position he died in, *"sitting up in his daytime clothes reading some papers he held upright in his hands."* It has convinced the world *"he was able to retain the papers upright in his hands in the midst of the immense pain of a massive heart attack due to the grace of God."* 4

It is a medical fact, because his death was unwitnessed and there was no evidence of trauma, only an autopsy could have determined the cause of death. Because the Pontiff died within Vatican walls, not Italian ground, the death was not covered by Italian law and the Luciani family had no authority to demand an autopsy. Yet, several field cardinals led by Cardinal Benelli of Florence demanded an autopsy, but one was never performed. Today it is a matter-of-absolute-fact no one knows what killed John Paul, except those who were responsible for his death.

[1] pg 184 Robert Hutchinson's *Their Kingdom Come*, pg 149
[2] pg 184 *L'Osservatore Romano* 3 Aug 78
[3] pg 187 Seabeck's *The Smiling Pope*
[4] pg 188 *L'Osservatore Romano* 9 Oct 78 Vatican corrected released circumstances of John Paul's death

Chapter 15

Pauper who would be Pope

He had come onto the world's stage, together with his secretary Lorenzi, in an outdated Lancia 2000. Had the automobile been brand spanking new, and it was not, it would have been totally unbecoming a common priest, much less a cardinal of the Church. Its fenders had been scorched by time and for the most part it had lost its color. So much so the Swiss guards stopped it at the Vatican gates and demanded identification. It was a tin box that had been designed and built for paupers. Forty days later he left in a pine box. A pine box designed and built for paupers. Two hundred and sixty-two Popes and ten thousand cardinals before him interred in grand, ceremonial, jewel encrusted caskets of gold and bronze and marble and he in a wooden box, this *Pauper who would be Pope.*

Funeral John Paul I October 4, 1978

He had made the rounds, pauper, then altar boy, then priest, then pastor, then monsignor, then bishop, then archbishop, then cardinal, then pope; then pauper, once more. He left a total estate valued at less than five thousand dollars, a checking account of about eleven hundred dollars, his Lancia 2000, some personal items including his personal clothes and two cockatiels. There was a pair of cut glass cruets which had been given to him by his mother on the day of his ordination. There were a few other gifts and medals and plaques that he had received through the years on various occasions. He had given everything he had to the needy, this *Pauper who would be Pope*.

Baby Pigeons

Nevertheless, I have some answers for you here. Among them are the baby pigeons. Baby pigeons? Yes, why there are no baby pigeons? One knows that dogs come from puppies and cats come from kittens and cows come from calves and sheep come from lambs and chickens come from chicks and ducks come from ducklings. But, just where do pigeons come from?

Or perhaps you have never noticed? You never wondered why in all of your life you have never seen a baby pigeon? Go to Saint Peter's Square in Rome or for that matter any of the grand piazzas in Europe or the great parks of America and you will see tens of thousands of pigeons and not a baby pigeon among them.

What's that? You believe they must be somewhere else? Well, go and try and find them and you will find that they are not there. I will prove that they are not there. For pigeons, among all of God's creation, come from somewhere else.

So I have some answers for you here. Among them is the answer to the greatest mystery of all. Something that, like the baby pigeons, you always assumed was there. Something that from day to day you could not see, yet, nevertheless always assumed was there. As in the case of the pigeons, I will prove that it is not there.

Now let us go back to that time, so very long ago, when I first visited the remote mountain province of Vittorio Veneto. Back to that time, when I first met a man called 'Piccolo'.

Chapter 16

Milan

The spring of 1969

It was an exact point in time, the first day of spring; that day on which the sun rests directly over the equator; that day on which all over creation the sun rises due east and sets due west and day and night end in a dead heat in time.

The Boeing 707 rose slowly out of Kennedy and started out over New York harbor before banking to the left and heading in its intended direction. As it made its turn, I looked down at that grand lady who lifts her lamp by the golden door and wondered how she ever came about.

After all, we were a Christian nation, our forefathers; every one of them was a Christian. Of those who had signed the Declaration of Independence, there wasn't a single black, not a single Jew, not a single gay, not even a woman among them. At least I didn't think so. It seemed to me a towering figure of Christ would be more appropriate here at these great gates to what was, indeed, the Promised Land.

In all of Europe I knew only one person. That is, I probably knew others but I didn't know where they were. But this particular one I had always kept in touch with. He had been one of my rivals back in high school in my run for the roses, one of those people I just had to beat out in life. His name was Jack Champney. He was much smarter than I was. His problem was that he didn't know it. He thought that I was much smarter than he was. He had gathered this opinion from how well I had done in high school; that I could keep pace with him all the way down the stretch only to lose him at the wire. He had no idea how much harder I had to work for what I got than he did. For him winning the race was like Frank Sinatra singing *My Way,* almost effortless.

He won the race going away. I was just another student in the crowd of several thousand when I watched him take the stand on graduation day. On this day, fifteen years later, I couldn't remember a single thing that he had said. And vainly I thought, "If it had been me, today, all that had been privileged to have listened would remember everything, every single word I said."

As the plane started to take down the time zones, I thought back to graduation day,

... that day they were all wearing blue blazers. Usually it was only me. Day after day, year after year, it had only been me. The others in sweaters, blue jeans, sneakers, whatever they laid their hands on when they got up in the morning. They used to call me 'pretty boy' - 'momma's boy.' Then one day one of them called me a 'pansy.' That's the day they found out my small fists could hit and my feet could kick. And sooner or later they learned my help with their homework made the difference between honors and failure. So, although I had made a run for it, I didn't come in first or even second, but, nevertheless, when the wreaths were passed out I took home with me the ones that were labeled Catechism, Mathematics and History. What's more, the yearbook caption alongside my picture read, "Most likely to become a cardinal of the Church."

Well, I was never to become a cardinal of the Church, for when the 'V' in the road came up, Jack took the path that said *Christ* and I took the one that said *Money*. I went off to the world of business where with him out of the way I would have less trouble reaping the roses.

He went on to attend Holy Cross where he once again took the honors and then began to work toward becoming a very special kind of priest. He took his doctorate in psychiatry at Johns Hopkins and registered as a licensed psychiatrist. His aim, made clear in his letters to me, was to become a member of the Catholic Church's Commission on Spiritual Occurrences. The commission, overseen by a panel of bishops, was the Vatican's investigative unit for apparitions, exorcisms, miracles, and other spiritual claims or happenings. He had secured an assignment as an instructor in a seminary and had the additional duties of secretary to the local bishop. The bishop happened to be a member of this commission; a nice stepping stone for Jack on his way to the Vatican. The cathedral, together with the bishop's residence, was located in the remote town of Vittorio Veneto in the foothills of the Italian Alps.

Just two weeks later, I had found my way across all of Europe to Italy. During that time, I had witnessed the splendor of Christianity. As I crossed Europe, I must have set the world record for visiting churches, including most of Europe's largest. Because it was too far out of my way, I had to skip Seville. But the others, all the others, from the majestic dome of St. Paul's; the great stone claws grasping at Notre Dame; the magnificent leaded glass windows of Chartres; the ashen wedding cake towers of Cologne; and now, finally, the great Cathedral of Milan with its threatening weather beaten gargoyles, oxidized by time, guarding the great square which lay before it; Milan's playground of princes and paupers.

I marveled them all. I relished them all. Yet it wasn't these great edifices that impressed me the most. It was something else, something I would find here in Milan, something I would stumble onto quite by accident; something I would carry with me all of the days of my life. Besides its great cathedral and its great square and it crystal galleria, Milan has a fourth great treasure, something most tourists never hear about; the city's great park of the dead. There you can see it all, every bit of yesterday.

The land of the dead

As I entered the cemetery, I became a part of it all. All was quiet. The only sound my footsteps and, perhaps, the breath of a slight breeze. The sky was foreboding as if it were a good day for a funeral. Had there been a lake, and there was none, it would have held dark waters as there was not a thing in the sky to give it life. It seemed all of the living had forgotten all of the dead. I was alone, alone as one could be.

Soon I realized I was wandering in the world's greatest metropolis of the departed; endless rows of mansions of marble, bronze, glass and some even of gold. There seemed to be more marble and granite houses in this land of the dead than there were houses outside in the land of the living; each one different as to command the attention of its own artist, its own architect, its own engineer.

Interspersed, here and there, were sculptures, mostly of marble and granite, yet a few of some precious metal and even one or two studded with jewels; each arrangement frozen in common death echoing its individual message of life; each one befitting a prince - no, a king - no, a god.

Collectively they echoed of immense wealth. "No wonder they lost the war," I thought, "they had all their money tied up in monuments." Nowhere could a single flower be found, as if to insure the beauty of God's creation not overpower these great works of man. There was only the green grass that worked its way like a maze in and out and around and about these great dwellings and monuments of the dead.

each arrangement frozen in common death echoing its individual message of life

Finally, I turned a corner and proceeded down what appeared to be the main boulevard of this great city of death. Flanked on either side by mausoleums of superlative grandeur, some sealed up like the tombs they were and many others showing off their merchandise. Through ornate iron grates, one could glimpse the sarcophagi themselves; mostly of marble, some of bronze, and a few of gold and even one of glass. And silence, silence all around, as not to wake those who were sleeping there.

As I came to the end of the avenue, I turned the corner and suddenly stopped dead in my tracks. Not dead dead, but dead in my tracks. There, just to my left, were a half dozen simple, tiny, matching white granite stones set in a row on a blanket of green grass which lay before a matching manicured hedge of green shrubbery.

The power of their simplicity eclipsed the grandeur of all that was around them. On each stone was carved a heart and within each heart a likeness of George Washington. "The Purple Heart," I thought to myself, and I thought something else, "Here is the real reason why they had lost the war."

Approaching closer with all the solemnity the moment commanded, in my mind echoed the faint sound of the bugle, followed by the hallowed roll of the drums and the distant roar of the cannon. A spot of light peeked through the overcast sky as if to mark this precious moment in time. As to give one light to read,

Frank Phillips, Sergeant
1921-1944
7th Army, 1st Battalion, A Company
Distinguished Service Cross

Richard Edwards, PFC
1925-1944
7th Army, 1st Battalion, A Company
Bronze Star

Jerome Rosenberg, 2nd Lt.
1919-1944
7th Army, 4th Battalion, A Company
Silver Star

Brian Pickering, Pvt.
1924-1944
7th Army, 3rd Battalion, B Company
Bronze Star

Anthony Jackson, Corporal
1922-1944
7th Army, 1st Battalion, A Company
Bronze Star

Patricia Wilde, 1st Lt.
1919-1944
Army Medical Corps
Bronze Star

It didn't say it, but I clearly heard it, "That they shall not have died in vain." One of those things one calls tears, crept up out of my heart and ran from the corner of my eye and moved toward its lid. I looked first to the right, and then to the left, and then again to the right, and finally to the left, once more.

Holding the tear on the edge of my lid, I said to them, speaking as if I were a great orator of some kind, "Not Thomas Paine with his pen, nor Patrick Henry with his eloquence, nor Paul Revere with his horse, nor Washington and Jefferson with all their courage, not even Lincoln at Gettysburg have spoken louder. For you have made more noise for freedom than all the others who have gone before you or have come after you. Believe me, I pledge to you this day, to each and every one of you, that each and every one of you will not have died in vain."

I have carried that pledge, that sacred duty with me all of my life. I have carried it every day, every hour, every moment of my life. I have carried it in my mind, and in my heart, and in my being, and in my very soul.

Now it is time to carry out that solemn promise, to answer that fervent prayer. To carry it out for each and every one of them; that what they dreamed of, those things they willed to be, will come to be, for each of them, and for me, and for you, and for all humanity!

Chapter 17

Vittorio Veneto

The next morning I took the train to Venice and it was from there I took the train to Vittorio Veneto. Exiting the station, I did exactly what Jack had told me to do. I took the first right, and then the first left, and again the first right, and finally, the first left, once more.

I stood in front of what appeared to be an old southern hotel; a southern hotel in the most northern part of Italy. It was surrounded by an aging wall eight or nine feet high and was of the same light shade of amber stucco as had been most every other building I had passed on the way. Like all the others, it was topped off by one of those orange terra cotta tile roofs that sprawl over so many Italian villages.

I followed Jack's instructions and pounded with both my fists as loud and as heavy on the great wooden door as I possibly could. Off in the distance I heard footsteps on wooden steps, then they seemed to have reached firmness for a time, then again on wooden steps, and then firmness, once more.

There was the juggling and clattering of the unlocking of the door and standing before me was a little old lady who looked as if she had stepped out of an Italian motion picture.

"You must be Lucien. Jack has told me all about you. We have a special place for you." I followed her into a small reception area where I scratched out a registration card and surrendered my passport. I asked her for directions to the bishop's castle as she led me to my room which window overlooked a canal that ran behind the hotel. I freshened up a bit and headed back out onto the street.

As I made my way from the hotel, I passed over a small ancient stone bridge and found myself channeled down a long narrow street hemmed in on both sides by stucco houses of all different shapes and sizes. The stucco was, again, mostly of the same amber color and was in a general state of disrepair.

It was as if I was to see a green house or a blue house or a yellow house, I would remember it all of my days. As I looked up I could see the edges of the roofs framing the blue sky, every one of deep orange terra cotta tile. On the town's edge I entered a tunnel that had been carved into an ancient fortress, and at the end of the tunnel I stepped out into medieval times, an ancient twelfth century plaza. I sensed Romeo and Juliet might be interred nearby.

Unlike the town, where stucco had been the tradesman's craft, here the buildings were entirely built of large blocks of stone. As a matter-of-fact, all of the ground was stone. The stone had been bleached through time to a stark white. The tatter and torn and wear of the ages had gone unrepaired and as the woman in the hotel had told me, the cathedral overpowered the plaza which lay before it. Kitty corner to its left, partway up a mountain, was what I rightly presumed to be the bishop's castle; a medieval group of turret-topped towers; only the tallest of which survived intact.

As I made my way toward the cathedral, I passed a large fountain built into the wall to my left, a giant stone trough with lions and gargoyles strewing water into the pool in front of them. There were a few statues of men and women, here and there, set into arched niches in the surrounding ancient buildings. There were many more arches that through time had lost their inhabitants.

The bishop's castle (upper left)   Arched niches along the way

On all sides of everything I could see, rose the great Dolomite Mountains. Jagged cliffs engulfed the town like a giant horseshoe; a bit of the medieval ages trapped in a rocky gorge. Midway into the plaza, I noticed a line of large, yellowing stucco buildings of more recent vintage down a street to one side. "Some kind of a college," I thought to myself.

Suddenly, out of nowhere, a man came running toward me wearing a tank top and running shorts and he both smiled and waved as he passed. He wore a large colorful caricature of Minnie Mouse on his chest. I sensed he must have been about fifty. I was surprised the recent jogging fad in the states had reached this remote part of Europe so quickly.

At the same instant, great bells rang out. Not in a rhythmic sort of way, but in a clanging sort of way, as if they did not want anyone to know what they had to say. The old church was of the identical washed-out color of its surroundings. Its nave was out of balance with its tower, as if they had run out of money in the middle ages, when I guessed the building had been built.

The steps told me something else. They started a hundred feet from the church and were as wide as the church itself and tapered up a flight of dozens of steps to the massive wooden doors of the great edifice. I looked for the wheelchair ramp and saw none. I thought of Moses' instruction in *Leviticus,* *"Whosoever he be of thy seed that hath any blemish, a blind man, or a lame man, or anything superfluous shall not approach the altar of his God."* I thought, "They sure followed his orders in those days."

I started my accent up a cobblestone road which was hemmed in by ten-foot walls on either side. It ran up the side of the mountain toward the bishop's castle. I climbed for at least an hour before I reached the castle gates.

Exhausted, I entered through what appeared to be the original castle entrance into a small courtyard. A fountain, not unlike several I had passed on the way, was set in the center of the courtyard. Water was splashing out over its edges onto the courtyard floor and was running down toward and around me and out of the castle gates. Off to one side was an old beat up car of forties vintage. The bishop's house, itself, was of beige stucco and was set within the castle ruins. Its focal point was a grand symmetrical staircase, one set of stairs leading up to a landing from the left and another leading up to the same landing from the right. Jack stood there smiling. I decided to take the stairs on the left.

Author's collection

The bishop's castle at Vittorio Veneto hemmed in by castle ruins

We got the usual "hellos" and the "boy, you don't look a day older" out of the way quickly and I said, "I see you don't have an age limit on runners here. I almost got run over down there." I mentioned the older man who had run past me in the church plaza.

Jack replied, "That was the boss, he turns in a few miles every morning. He was brought up on the other side of the tracks and damn near starved to death as a kid. Having been a frail child, he developed himself into quite an athlete as a teenager. He still keeps himself in very good shape. He enters every 10K that comes up and usually wins in his class.

"If you had showed up a couple of weeks from now instead of today, he would take you up a mountain or two with him. He's an accomplished mountain climber and holds the speed record for several peaks in Northern Italy.

"When he comes back, I'll introduce you to him, but you won't see much of him until dinnertime, as he has to go to Venice today. Let's go inside. I'll show you my little corner of the world." He pulled back the large wooden door; its opaque glass panels protected by heavy iron lattice work required the strength of both arms.

We entered a large open space, the reception area of the house. Definitive paths had been worn into the ancient stone floor. One could make out precisely where people had walked through the ages. There was nothing there at all. That is, there wasn't a single piece of furniture; just open space leading to colorless, leaded glass windows at its far end which I correctly guessed overlooked the village of Vittorio Veneto.

Today, this room is an impressive introduction to the house; its walls are now lined with beautifully framed life sized oil portraits of the dozen or so bishops who have lived there in the twentieth century. At the far end of the room, the portrait of Albino Luciani, the only bishop of the Veneto country to have risen to the papacy, is hidden behind a door that leads to a small chapel on the left, what is probably best described as a prayer station. Like other portraits and statues of Papa Luciani that have been commissioned by the Church which are scattered across northern Italy, from the small village he grew up in to Venice, this portrait depicts a man of one hundred and five who is in the final days of what has undoubtedly been a long, unsuccessful bout with cancer; a part of the Vatican's ongoing deception plot to convince the public that this man was at death's door when he was elected pope.

The hand painted arched ceiling in this grand vestibule contains the record of more than eighty bishops who have served in Vittorio Veneto since the eighth century, each one represented by his coat of arms and period of reign.

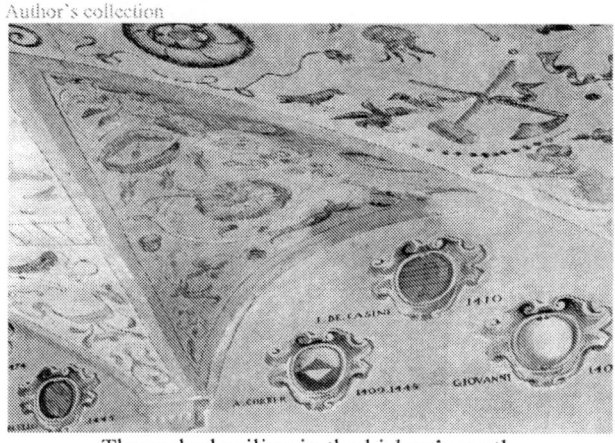

The arched ceiling in the bishop's castle

The trappings of the room I speak of in what is to follow as the bishop's office have since been moved to a monastery. Today, off to the right of the main reception hall is a succession of four large rooms of equal size, each leading to the next. The ceilings are paneled with richly carved squares of mahogany and the flooring is a mottled marble. The modernization has left the walls stark white.

The first of these is a large reception room furnished in rich Italian provincial furniture with conversation areas at each end. This reception room leads in turn to the secretary's office which today is quite modern in appearance, including a copy machine off to one side and a computer set on a modern desk. This room in turn leads to another reception room, again outfitted with Italian provincial furniture, which in turn leads to the bishop's office.

There is one piece of furniture in the house that is a focal point of history, a small glass fronted curio breakfront that contains a few of the Pope's personal items, the only record of his life other than his tomb that remains in the Church today; a small prayer book, his rosary beads, his red bishop's cap, and a few other items.

In a recent visit to Vittorio Veneto in the winter of 2005, the good nun who now cares for the house unlocked the case for me. I was able to touch some of the objects that were so close to him. On the bottom shelf was his bishop's ring and right next to it was Jack's class ring; a duplicate of one that I had in an old box at home. I picked it up and told the nun of its origin. She offered it to me, but I thought best I leave it where it belonged.

So there have been some changes as the house survives today; yet at the time of my first visit, now almost four decades ago, except for the arched painted ceiling in the main reception area, things were quite different.

"Blessed are the poor . . . for theirs is the Kingdom of Heaven"

Jack took me to the far end of the entrance hall and we accessed a small alcove, his little corner of the world. A very little corner it was. The office was tiny. I visualized my own office back in the states overlooking the marina at the corporate headquarters of the company in which I was rapidly making my way up the ladder. "I was certainly winning this race," I thought to myself.

The windowless room was as bare as it was small and clashed with its rich green, black and white marble terrazzo flooring, which instead of giving it the feeling of wealth which it reflected, gave it a feeling of coldness. It was quite obvious the flooring here was not original; it had been added to the house in more recent times. Most of the opposite wall was made up by the room's centerpiece, a beautifully carved mahogany door set in an archway, which I presumed to be the door to the bishop's office.

The unbroken line of the yellowing wall was interrupted only by this door and a small bargain basement crucifix above it and a cheaply framed photograph of Pope John XXIII on the wall behind the desk. The desk itself was more of an old wooden slab on legs than it was a desk. On the opposite wall hung one of those cheap alpine clocks one picks up in souvenir shops.

A message was written on the photograph. I approached it and found it was written in Italian. I heard over my shoulder,

"Albino Luciani,

Christ has asked me to express His very personal congratulations on this very important day of your life."

John XXIII
27 Dec 58"

Nodding at the richly carved mahogany door, Jack said, "It was given to the man who occupies the adjoining cell when he was raised to the rank of bishop. Piccolo was the first bishop installed by John XXIII. The ceremony took place in St. Peter's."

I was struck by Jack's reference to the bishop as 'Piccolo' and, at the same time, by the yodeling of a little nun who ran out of the alpine clock on the opposite wall. Sensing my shock at the sparseness of his little corner of the world he laughed, *"Blessed are the poor . . . for theirs is the Kingdom of Heaven."*

"When John first told him he intended to elevate him to the office of bishop, Piccolo objected citing the fact he was not an accomplished speaker and would be an embarrassment to Mother Church. Actually, although he had no problem speaking to his own congregation and youth groups, he possessed a fear that he might freeze up before important audiences.

"Although Piccolo did not tell him this, John surmised it. Not letting on that he was aware of this great hurdle which Piccolo felt he might not be able to jump, John started working secretly on Piccolo's training program. Working through intermediaries, he set him up on a series of increasingly important speaking assignments in small parishes, and finally he tricked him into his big test."

"His big test?" I repeated his statement in a question.

"Yes, his big test. Cardinal Colombo, the Archbishop of Milan, was to address members of the *College of Cardinals* and other Church dignitaries in the Sistine Chapel in Rome. The cardinal called Piccolo to Milan, where he had him write a speech for him. The address was to be heavily steeped in nineteenth century theology, something in which Piccolo excelled.

"Colombo then asked Piccolo to accompany him to Rome to help guide him in his rehearsals for his upcoming address. On the day he was to speak, the cardinal suddenly fell ill and Piccolo, being the only logical understudy, was faced with three options: he could resign the priesthood, he could feign illness, or he could try.

"The boss told me it was the most difficult decision he was to make up to that point in his life. He thought back to his father's instruction, '*You must find the strength along the way to do what you have to do to bring about a day when all children of God accept each other as equals.*'

"Piccolo told me he realized he had been set up when halfway through his address he saw John, who was sitting in a darkened corner of the chapel, wink his approval to the cardinal who sat beside him, Cardinal Colombo; the one he had replaced.

"'John was the angel who took me to the pinnacle and taught me to fly. I remember, clearly,' Piccolo recalls, 'as if it were yesterday … *the angel took me to the pinnacle of the mountain, and like a father condor teaching his baby chick to fly, commanded that I COME TO THE EDGE OF THE CLIFF, and I refused saying "I'll fall"; and then he repeated his command in a louder voice COME TO THE EDGE and I again responding in a weaker voice "No, I'll fall"; and finally in a roar of thunder he repeated COME TO THE EDGE, and with all the courage I could possible muster, I went to the edge, and he pushed me, and I flew.*'

"After he came down from the podium, John approached him and asked, 'Do I now have your permission to elevate you to your ecclesiastical destiny? Do I now have your permission to make it possible that you can better make the contribution that Christ requires of you?'"

"Piccolo was installed as Bishop of Vittorio Veneto the following week. His first words to his newly acquired congregation recalled his impoverished days as a child, *'Christ picked me up from the mud in the street and gave me to you.'* His will requires that these words be carved on the plain pine box in which he will someday be entombed.

"Actually, Piccolo told me he didn't give the speech he had written for the cardinal at all. Instead he told them, *'Our great enemy is the bigot who lives within each one of us. His ally is scripture. Our ally is conscience...'"*

I quickly cut him off, *"'... His ally is Moses. Our ally is Christ. As long as he exists no man is truly free... Believe me all the bigot has are words, flimsy words, Flimsy arrows that he thinks will win for him. Flimsy arrows built in an ancient factory.'"*

Jack who had been stopped dead in his tracks by my dissertation exclaimed, "Good, you've done your homework. It will help you in your mission."

"My mission?" I questioned.

"We will get to that later. Yet, this was a big turning point in his life. As a priest, he had been limited to accomplishing his objectives through the press. As a bishop, he has become a rampaging locomotive running about the courts and legislative halls of Italy in what has been an endless struggle to achieve equal rights for all."

Mostly because of his toes

We continued to chat for awhile and he had just told me Piccolo must have run off to Venice already, "... but you will meet him at dinner tonight. He has been looking forward to meeting you," when suddenly, at the outer doorway stood the man who had run past me in the park. He was clad in an Hawaiian shirt, shorts and sandals. His wet hair told me he had just come out of the shower.

Two things struck me, the countenance of his smile, and the other was his toes. I never thought of bishops as having toes, especially toes that were as perfectly pedicured as these were. I wondered if the nearby convent provided this service for free to bishops, a service that when I splurged for it cost me fifty bucks a throw back in the states. Yes, a third thing struck me - his voice, as he introduced himself as the man next door.

The voice was a piping, rasping voice. Not as if he was talking through his nose, but as if the pipe was built into his throat.
It was a one-of-a-kind voice. It was this that made it easy for me to follow what otherwise would have been heavily Italian accented and somewhat broken English.

Caught off guard, I had forgotten the title one uses in addressing a bishop of the Church. I stammered, "I apologize, I forget what I'm supposed to call you?" The bishop cut me off, "Just call me the same thing everyone else calls me, 'Piccolo'."

To this day I have never forgotten Piccolo, partially because of his one-of-a-kind voice and unrelenting and charismatic smile, but mostly because of his toes.

After a brief chat he said, "I must go, for the thief of the ages is knocking at my door."

I turned toward Jack, "Did you hear someone knock?"

The Bishop of Vittorio Veneto laughed,

"Time. Time is knocking. Time is a thief. For it will rob one of one's childhood, eventually deprive one of one's youth and ultimately take one's life. But it is a good thief. For it provides the span of wonderment for the child, the term of enlightenment while he grows and the age of fulfillment as he gives..."[1]

He added, "And my time to give is near." and he was off.

"Kind of a piercing voice, huh?" Jack offered, noticing that I had reacted to the bishop's voice with a start. "Actually it was at the ground roots of his fear of public speaking. When he was a teenager, Piccolo had a tonsillectomy that went haywire and left him with his uniquely thin, raspy voice; a handicap which John, as I just mentioned, cured with one of his miracles."

"Blessed are the poor in spirit..."

"Anyway," he said getting up from his chair, "I might as well show you where the boss spends his time dreaming up the next chapter of events for this sprawling paradise here in the foothills of the Italian Alps." Rising, he moved toward the great mahogany door and opening it, we proceeded in.

The office could not have been more impressive. My heart sank a bit as I thought of my relatively modest surroundings back in the states. I also thought of the tax-free exemption status of the Church. Yes, the rich marble terrazzo flooring continued into the room, but here it was not out of place. The walls were mahogany; that is, ornately carved mahogany edged in gold. In each of what must have been three dozen panels which surrounded the room stood a life-sized beautifully painted angel in Byzantine fashion. Each one, unique in color and dress, sported a halo of golden leaf. Yes, each one was armed with a formidable weapon.

In addition to their protective presence they seemed to be listening, as if all that would be said within these walls would be related to the one above; kind of doing the same job here as the tapes were doing in the Oval Office some four thousand miles away. Two more life-sized angels, these in three dimensions and of white marble, guarded the huge walk-in fireplace.

Just above the mantel was mounted what was obviously meant to be the focal point of the room - an old well-weathered oil painting of Christ driving the moneylenders out of His Father's temple. Its dark tones were accented by the brilliance of golden coins which cascaded from the tables and seemed to splash down out of the painting itself, out beyond its overdone, heavily encrusted, golden frame. I thought it could be a Rembrandt.

Jack, noticing I was appraising the painting, offered, "No, not Rembrandt. It's by Titian. This is Titian country. Some of his best works are in Vittorio Veneto. His masterpiece is the altarpiece in the cathedral you passed on the way. Piccolo put this one here to remind us that Christ too had to deal with the republicans. He says this is proof, proof in the scriptures themselves, the republicans trace their roots back to Christ; that is to the time of Christ."

The room was divided down the middle by a low altar railing with twisted golden columns built into a rich mahogany framework. On one side of the room was kind of a boardroom with a huge coffin shaped mahogany table edged with gargoyles and cherubs. Two rather wide Persian carpet runners lay along each side of the table. They were of a kind so plush that if one didn't take notice one would easily trip over them. On them sat a dozen richly embroidered chairs with matching mahogany gargoyles shooting out of their arms. There were six on each side of the table. There were no chairs at the table's ends, as if one end was reserved for the invisible presence of Christ and the other end reserved for God the Father. As I looked up, there was a beautifully hand painted ceiling with more angels in its arched niches.

On the other side of the great room was the bishop's personal work and reception area with an elegant sofa in Italian provincial, and a low matching coffee table and armchairs. A grand kidney-shaped, ornately carved desk with neatly arranged writing and blotting instruments sat before a stark red heavily brocaded wall. I took particular notice there were no papers on the desk. As a matter-of-fact the whole room seemed to be set for display in a museum rather than set up as a workplace.

On the wall just behind the desk, the only break in the perimeter walls of the room other than the door in which angels did not stand, were two relatively new life-sized oil paintings. The matching simplicity of their inexpensive modern frames clashed with the ornate antiqueness and wealth that was all about them; one of the reigning Pope, Paul VI; the other of his predecessor, John XXIII. Between the two paintings was a framed document, this one written in English,

Republic of Italy

"For extraordinary heroism while engaged in military battle without regard to his own safety and risk of life no matter how great, to the betterment of lives of others no matter how small; the Republic of Italy is forever indebted to our eternal friend.

Aldo Moro
March 29, 1965"

"Wow. he is a war hero." I exclaimed. "Not exactly, Jack told me. I will tell you about that later." With an athletic curl he rolled himself into the large leather chair.

Opposite were two pairs of rather large French windows of relatively recent vintage which overlooked the courtyard outside. Suddenly, the somber setting came to life with the chiming of the hour. Off to one side was a massive grandfather clock which reached upward toward the ceiling. The ceilings must have approached sixteen feet in height and glancing back at the door and then to the windows I wondered how one ever got a clock of such massive proportions into the room.

"It's really a great, great, great, great, great grandfather clock," Jack offered. It was designed and crafted by Valdini, an Italian craftsman living in Switzerland in the seventeenth century. Valdini's clocks normally fetch upwards of a quarter-million dollars; on occasion, a half-million.

"It was given to the diocese when Leo XIII visited here late in the nineteenth century to celebrate the establishment of the seminary. The rest of the contents of the room including its paneling, the ceiling and mantelpiece came out of a monastery that was set on a mountain outside of Naples. It was moved here to Vittorio Veneto as a protective measure when the allies reached Sicily and it had become imminent they would take Italy. The Pope thought the allies might destroy it. He was right as they bombed the hell out of it, thinking that it was being used as an Axis headquarters."

Summing up what I had just witnessed, I thought, "Blessed are the poor for theirs is the Kingdom of Heaven." I thought it best to keep the thought to myself.

As if he could read my mind he offered, *"'Blessed are the poor... for theirs is the Kingdom of Heaven.'* Actually," he corrected himself, *"'Blessed are the poor in sprit for theirs is the Kingdom of Heaven,'* are Christ's actual words, the only time Christ is known to have selected a particular population group other than children and guarantee them paradise.

"Piccolo says it refers to those who reason, those who don't easily believe, the poor in spirit. Those that use their conscience," he pointed to his temple. "Those Piccolo calls the lions, lions as opposed to sheep. But the mass of theology prefers to believe

Christ was talking about the poor. Certainly, that is not what it says. Christ very obviously meant exactly what He said, *'the poor in spirit.'* Those lions who believe their search for the truth requires some effort of their own, rather than those sheep who just assume it is handed to them on a silver platter in their scriptures when they are born; their birthright is their ticket to salvation.

"Actually," he went on, "Piccolo rarely uses the office. He's never been a fan of the pomp and splendor of the Church. He just doesn't feel comfortable here. He handles his paperwork on the dining room table.

"Each of us sees in the great clock a work of fine art and scientific achievement. Piccolo sees something else. He sees in the clock the right to a good and healthy life for hundreds of children. Within a week of his arrival, Piccolo started to sell all of this and more in order to build an orphanage for children who would otherwise be aborted. But the Vatican intervened and stopped him.

"If you have followed him in the world press, he is a threat to the regency of Rome should he ever rise to the papacy. Should he ever become Pope, it is widely rumored his first act would be to leverage the Vatican treasures and use the money to annihilate poverty in third world countries. One reason why most people believe he will never rise to the top."

The adversary

He walked to the immense table. Taking a seat at one end he waved me into the chair opposite him. "A few words about our adversary," he began.

"Our adversary?" I questioned.

"Yes, our great adversary, Moses," he replied.

"Moses?" I questioned again.

"Yes, Moses," he repeated. "Moses is the quartermaster of all hatred in the western world. It is he who supplies the bigot with his arsenal of weapons. It is a bottomless pit, this arsenal that gives the bigot the words he needs to conduct his evil war.

"It is Moses in the Old Testament who tells us *'It is God the Father's command that woman live in dire servitude to man.'* It is Moses in the Old Testament who tells us *'It is God the Father's command that slavery be a way of life.'* It is Moses in the Old

Testament who tells us *'It is God the Father's command that those with flat noses, those who we call Negroes today, and the handicapped are subordinate peoples and are not worthy to approach the presence of the Lord.'* It is Moses in the Old Testament who tells us *'It is God the Father's command that born-out-of-wedlock children are not to approach the congregation of the Lord.'* It is Moses in the Old Testament who tells us *'It is God the Father's command that homosexuals are outcasts and are to be put to death.'* It is Moses in the Old Testament who tells us *'It is God the Father's command that sex is shameful and sinful.'* It is Moses in the Old Testament who tells us *'It is God the Father's command that the prostitute and the adulteress shall surely be taken outside the city and stoned to death.'* It is Moses in the Old Testament who tells us *'It is God the Father's command that whoever does not seek the Lord God of Israel shall surely be put to death.'* It is Moses in the Old Testament who tells us *'It is God the Father's command that only the sons of Aaron, the Aryan race, are to serve at His side, that all others are to be annihilated or cast into slavery.'*

"It was Moses who, through his commanding general Joshua, carried out God the Father's horrific instruction in the taking of the Promised Land in which hundreds of thousands of helpless men, women, children and infants were slaughtered, *'Put to the sword the unbelievers, the infidels, every man woman and child, leave not one alive.'*

"It is Moses who gave birth to all ethnic cleansing atrocities since, including the terrible undertaking of the Crusades, the bloody American Civil War and most recently the Holocaust in which six million Jews lost their lives.

"Each of these instructions of the Bible and many more like them are prefixed with the words *'And God spoke to Moses.'* It is these words that make the Bible scripture. Without them, it would just be another book, just another fairytale. They tell us explicitly what kind of God we are dealing with in the God of Abraham and Moses, a God of hatred and evil." He raised his voice a notch or so to emphasize his final point, "No preacher can stand on a stage and tell his congregation otherwise, for his scripture is his adversary!"

Ethnic Cleansing

He paused again to give me time to grasp all the terrible things he had said. He continued, "The validity of Moses depends entirely upon whether or not he was telling the truth. Was Moses telling the truth when he told the story of the four hundred years the Israelites spent in Egypt? Was he telling the truth when he told the story of Joseph and his brothers? Was he telling the truth when he told the story of the Ten Plagues? Was he telling the truth when he told the story of the Exodus? Was he telling the truth when he told the story of the forty year wandering in the Sinai Desert? Was he telling the truth when he told his people it was God the Father who appeared to him on Mount Sinai and had given him the Ten Commandments? And, finally, was he telling the truth when he told the stories of Creation and of Adam and Eve and of Noah's Ark?

"For if Moses was telling the truth, then the bigot is right; bigotry would be a virtue. As a matter-of-fact it would be the only path to heaven, for to annihilate all the others is the only way that the Aryan race would end up representing all of mankind. It is the only way in which God the Father's dream as told by Moses in the

Old Testament could come true. But on the other hand, if Moses was lying, then bigotry is an abomination, the abomination all men or women of good conscience know it to be. So the bottom line is, was Moses telling the truth, or was he lying when he told his many stories?

"This is one of the great hurdles we face. The mass of the Christian world perceives Moses to be some kind of saint, a holy man of some sort. This is the great phenomenon of faith. This is because the picture books and the motion pictures influenced by Christian preachers have created the illusion Moses was a holy man. Yet, anyone short of an imbecile who reads the five books of Moses can only come up with the conclusion that Moses was a monster."

He raised his voice, this time an octave or so with a tinge of frustrated anger, "There was nothing holy about this man at all. He has led the world into thirty-three hundred years of hatred and prejudice and persecution and horror and suffering and destruction and death!"

Silence prevailed and I finally broke the stillness, "Well, it is quite clear that no one could ever prove Moses was not telling the truth in what he had to say. After all, none of us were there; we don't really know what God the Father told him."

Suddenly Jack stopped. He peered around the room, carefully examining the expressions of each of the angels which surrounded us. His action was quite decisive and it gave me the chance to count them, thirty-three in all. I thought of the thirty-three years Christ had lived and of the thirty-three months of His ministry and then of the thirty-three months Anne Frank had hid in the attic. In retrospect, today, I could add the thirty-three days of Luciani's papacy.

On the other side of the aisle, I thought of the thirty-three centuries Moses had wreaked havoc and bloodshed upon mankind. Jack gave me a look of uncertainty. He opened a drawer in the table and withdrew a couple sheets of paper and said, "Let's go for a walk."

---

[1] pg 206 Luciani originally said this in eulogy to John XXIII in the Basilica de San Marco in Venice 4 Jun 63

Chapter 18

The Luciani Expedition into Egypt

As we went out of the castle gates, a light rain was falling. Not enough to justify an umbrella, but nevertheless it was there, a light mist.

"I am about to show you our best kept secret." I followed Jack along a path which wound down around one side of the snow peaked mountain. Shortly, we found ourselves in the rocky gorge in which the village was trapped. Making our way through a mile or so of lush underbrush we came into a grotto enclosed by sheer walls of darkened slate-like rock on three sides several hundred feet high. Cascading off its center panel was a waterfall. It was splattering downward onto a bed of rock and then into a pond that was edged in with splashes of green reeds.

As I stood there I thought of all earth's species, for I could hear the hum of bees, the chirping of birds, the croaking of frogs, and out from under a bush and hesitating, as if to size us up, was a snowshoe rabbit. I could clearly see a number of fish of many different shapes and sizes glistening in a wide range of colors of silver to gold to blue to red; a living rainbow making its way in the crystal clear water before us.

"Piccolo and I call it our Garden of Eden," Jack said, "We do our best philosophizing; our best thinking here. And," he added, looking about him suspiciously, "we tell or best secrets here."

As I looked around I began to realize Adam and Eve had been given a lot to work with. A silent wind crept about the tips of the evergreens clustered just to the right of the waterfall. They whispered spring's early return. A huge umbrella palm stood just to the left of the waterfall and under its protection the rocks, which edged the pond on the opposite side, were weed-free. Where the weeds should have been, lilies of the valley sprouted here and there and everywhere.

Next came a range of small trees; half circling the pond, they formed a crescent so as to reflect the moon which was not there. Just behind them was a row of taller trees; so overpowering they seemed to imprison the tiny crescent before them. As the mist made its retreat, tiny slivers of the sun's rays pierced through them as if trying to take out a frog or two which sat on island rocks here and there in the pond. Altogether, like a family reunion posing for a picture.

On our side of the pond was spreading undergrowth, much of it flowering with red, white, yellow, blue, orange; whatever colors the Master happened to have dabbed into with His paintbrush when He executed this breathtaking work of art. Some of the color was in motion; butterflies and fireflies hovered here and there and everywhere.

Author's collection

flowers, butterflies and island rocks

Just to our left, yes, I couldn't believe it, was a small tree wearing its spring garb, apple blossoms. Angel winged white petals drifted down to the waiting waters. I thought, "Could this young tree be a descendent of the one that started it all?"

Seeing that I had noticed the tree, Jack laughed and said, "Piccolo planted it ten years ago when he first came here and just about everyone who sees it thinks the same thing you were just thinking. No," he corrected my thought, "its parents were American; it is an American apple tree.

"Piccolo has always been fond of Americans. He's proud of what they have done and are doing. He is particularly proud of the Kennedys and the work they have done for the handicapped, including having introduced the Special Olympics. Just last year Piccolo held the first Special Olympics ever held outside the United States here in Vittorio Veneto.

"Ten years ago when he first arrived here, Piccolo attended the local festival and was disturbed that not a single severely handicapped child was to be seen. Shortly afterwards, he set up a clinic, a kind of halfway-house, designed to take many of these children out of institutions and allowing them to live together with the general population.

Once again Moses stood in his way. Many mothers and fathers didn't want their children exposed to children who were severely impaired. They even went so far as to picket the workers when they were building the first ramp into a church here in the diocese, actually the first one built in the world. Their picket signs went so far as to quote the words of Moses in the Old Testament.

"Actually, President Johnson is on top of his list for having fought off the *Christian right* to bring about human rights for blacks. Piccolo knows that the future of the world depends largely on what Americans do. That's one reason I'm here. Of course there is another reason why I'm here."

"Another reason?" I asked.

"We'll get to that later," he answered.

Baby pigeons

Two large rocks sat at the pond's edge a conversation apart. He took one and I the other. He paused for a moment or two as to give me one last chance to take in all of the wonderment that was about us. Then he started, "The angels in the castle most likely do have ears and it is for that reason I have brought you here. Although we have never been able to find them, Piccolo and I believe the house is bugged."

He paused a moment and then started up again, "Just before we left the office you remarked, 'No one could ever prove Moses lied in what he had to say, as none of us were there. We don't really know what God the Father told him.'

"Not entirely true. Piccolo came up with the idea a couple of years ago that if one could prove Moses was lying when he told his stories then one could rid the planet of bigotry once and for all. It would destroy the arsenal of weapons from which the bigot draws his ammunition. Piccolo decided to go after that part of the Bible where the historical benchmark is exact, the time of Moses.

"He determined the foundation of all that Moses had to say was his story of the four hundred years the Bible contends the Israelites were in Egypt, which marked the beginning of the Hebrew nation and the birth of Christianity. If the Israelites were actually in Egypt, as Moses claimed, then everything Moses said would fall into place. It would prove he was telling the truth. On the other hand, if one could prove they were never in Egypt, everything Moses said would fall to ruin.

"Piccolo, being a bishop of the Church, couldn't conduct an onsite investigation without Vatican approval. He proposed to the commission, if such an investigation were to prove the Israelites' occupation of Egypt, it would authenticate the word of Moses and Christianity. It certainly would, as you know, there is no other hard evidence that proves Christianity; it is simply belief.

"So, with Vatican approval, Piccolo and I and Brother Tom Jones, a local monk, set out for Egypt during the summer school break just last year. Our objective was to uncover evidence of the Israelites' presence in Egypt during the Middle Kingdom Period, the period 1750BC to 1350BC, the four hundred years the Bible contends the Israelites were in Egypt; in Egyptologist terms, the twelfth through the eighteenth dynasties.

"It took us almost two years to prepare for the expedition. We had to become knowledgeable in ancient symbols and several languages including Egyptian hieroglyphs and hieratics and Hebrew, Nubian, Hyksos and Greek script. For Piccolo, the job was much easier, as he was already a scholar in most of these languages. He was more of a teacher than he was a student.

"According to the Bible, it was because of their great numbers the Israelites were put into slavery during the last forty years of their alleged time in Egypt. *'Pharaoh said, Behold the children of the people of Israel are more and mightier than we: Come on, let us deal wisely with them less they turn against us. So with hard bondage of mortar and brick the Israelites built for Pharaoh the treasure cities of Pithom and Ramesses.'"*

"What did you find?" I asked, quite eager to learn of this link between the Bible and reality.

He paused for a moment as a great detective often pauses before revealing the solution to his case. "Nothing." he replied. "All we did was to verify the fact the Israelites had never been in Egypt.

Ancient cemeteries at Pithom and Ramesses

"We examined literally thousands of ancient tombs, about four hundred of which have been excavated in the ancient cemeteries of the cities of Pithom and Ramesses where the Bible says the Israelites were concentrated; the Exodus originates from this part of Egypt. There were many thousands of ancient grave markers in these cemeteries dating to 1750BC-1350BC.

"We couldn't find a single Hebrew marking on any grave. In fact, prior to our investigation, in all of Egypt only two or three Hebrew markings or names have ever been found in all of the Egyptian tombs and palaces and burial grounds, which are otherwise loaded with hieroglyphs. These few markings have always been accompanied by the Star of David which dates them several centuries after the Israelites presumed time in Egypt, as King David didn't come along until then. There is not a trace of Hebrew history in Egypt dating back to Moses time."

"Nothing?" I questioned, astonished at the revelation, for it was something the Christian world had always taken for granted. It would be like looking at a photograph of the Empire State Building after all these years and assuming all the time it had a first, second, third, fourth and fifth floor; floors that although concealed by low-lying buildings around it, were very obviously there. It is quite obvious the Empire State Building has a first, second, third, fourth and fifth floor as otherwise one could have never added the other ninety-seven floors.

Although no one has ever seen one it is quite obvious there are baby pigeons. Otherwise there would be no pigeons. It is quite obvious that Christianity has a first, second, third, fourth and fifth floor, the five books of Moses, as otherwise one could have never added the other sixty-two books of the Bible. One doesn't have to see it to believe it. It is quite obvious the Israelites were in Egypt. No normal mind would question this fact. I shuddered at the magnitude of what Jack was suggesting.

Jack answered my unasked question. "Yes, nothing at all. We found nothing at all. And there is much more to it than just the graves. There is the great volume of hieroglyphs in the tombs and palaces of Egypt, which if reduced to fine print would fill all of the volumes in the New York City Library and then some. There are over sixty thousand rooms in the tombs that have been opened so far and they are covered from floor to ceiling with hieroglyphs. In addition, there are tens of thousands of clay tablets and many temples and palaces with even more hieroglyphs.

"Also, about a third of this volume and the most complete record pertain to the time the Bible claims the Israelites were in Egypt. The hieroglyphics spell out in great detail every major event that occurred during this period of Egyptian dominance: the birth, period of reign and death of every pharaoh, every war that was waged, every battle that was fought, every celebration, every famine, every plague, every invention, every advancement in civilization; but nothing, not a single trace of any of Moses tales.

"Despite the Biblical 'fact' that more than half of its population was Israelite, there is not a single mention of the way of life of the Israelites in Egypt. In fact, no mention of the story of the Red Sea miracle in which Moses claimed the entire Egyptian army and its Pharaoh perished which if true would be the greatest event in all of Egyptian history; nothing, nothing at all, not a trace.

"In addition, the Cairo and British museums maintain a catalogue of the roughly twelve thousand mummies that have been excavated in the nineteenth and twentieth centuries. Included, is a complete description of each mummy - its time of interment, its age at death, its cause of death, its race, its nationality and so forth.

Christ's family tree

"This led us to another conclusion. The Old Testament is an exact genealogy from Adam to Christ. The Bible lists a direct chain of fifty-five generations between Adam and Christ's father Joseph."

He reached in his pocket and handed me a sheet of paper. I read to myself, *"Adam, Seth, Enos, Cainan, Mahalaleel, Jared, Enoch, Methuselah, Lamech, Noah, Shem, Arphaxad, Salah, Eber, Peleg, Reu, Serug, Terah, Abraham, Isaac, Jacob, Judas, Phares, Esrom, Aram, Aminadab, Naasson, Sakmon, Booz, Obed, Jesse, King Solomon, King David, Roboam, Abia, Josaphat, Joram, Ozias, Joatham, Achaz, Mahasses, Amon, Josias, Jechonias, Salathiel, Zorobable, Abiud, Eliakim, Azor, Sadoc, Achim, Eliud, Mattham, Jacob, Joseph the father of Christ."*

He told me more. "The chain is unbroken and the Bible explicitly gives the precise age at which all but three of them sired his firstborn. In addition, it specifically gives the age of death of each one of them. One knows who was whose father and who was whose son all the way from Adam to Christ. *'Adam begot Seth who begot Enos... who begot Mattham who begot Jacob who begot Joseph, the father of Christ.'*

"In ancient times, there were no social requirements restricting the age at which one could sire one's firstborn. The average generation length was under seventeen. The mathematical calculation 55 (generations) X's 17 (generation length) gives us 935 years. It is an historical fact Herod died in 4BC. This would place Christ's birth in 6BC. This would place the time of Adam at 941BC. Moses and the early prophets who followed him were obviously not very good at math.

"Bibles tell us something quite different. Those patriarchs who lived between 3100BC, the time of the first pharaoh, and 1400BC, the time of Moses, lived to an average age of 273 years; the oldest living to be 960 years. However, the average lifespan of the pharaohs, for the same period, was 36 and the oldest, Ramesses II, lived to the extraordinary age at that time of 86.

"Most scholars believe that when the stories of the Old Testament were first put together in a single volume, the authors realized, if these patriarchs had lived normal life spans, it would place the time of Adam and Eve sixteen hundred years after the great pyramids had been built. They added these extraordinary lifetimes to push the date back to 4000BC; acceptable then, but as one knows 4000BC makes no sense today. We all know mankind goes back hundreds of thousands of years beyond that.

"Nevertheless, getting back to the mummies. Although the more important ones are displayed in museums, many of them are stored in warehouses by the Egyptian and British governments and some of them have been returned to their tombs. About a quarter of them have been dated to the Middle Kingdom Period, the time the Israelites were said to have been in Egypt. The overwhelming number of them are Egyptian, there are a few Greeks, Hyksos and Nubians, but there is not a single Hebrew among them.

"This is consistent with the tombs themselves. In the tombs uncovered to date, there are more than three million painted and etched human figures depicting the way of life in Egypt for three thousand years. The great majority of these are Egyptian and there are many Greeks, Hyksos and Nubians among them, but there is not a single Hebrew depicted among them despite the fact the Bible claims more than half the population of Egypt was Hebrew for four hundred years.

Author's collection Cairo Museum

Over three million figures and not a Hebrew among them

"Keep in mind the Bible claims some Israelites attained royalty status including Joseph, himself, *'So Joseph died being an hundred and ten years old: and they embalmed him, and he was put in a coffin in Egypt.'* None of these tombs has ever been found.

"Today we know the name, the date of birth and death and the period of reign of every single queen and pharaoh who reigned in Egypt from 3100BC to Cleopatra, a generation before Christ's time. We have recovered the mummies of all but a few of them, including all who reigned during the Middle Kingdom Period. This includes the remains of Amenhotep, the pharaoh who is presumed by the Bible to have drowned in the Red Sea. They have all been Egyptians except for the Hyksos kings who ruled during their time and two who were Nubians. The chain is unbroken."

He clarified his statement, "Bibles dated prior to 1871 placed the Exodus in the reign of Ramesses II. In that year, he was found to be safe in his father's tomb in the Valley of the Kings. Bibles after that time place Exodus in the reign of Amenhotep II. When they found his tomb in 1886, the so called experts gave up.

Mummy of Ramesses II

Tomb of Amenhotep II

He reached into his pocket and pulled out one of the sheets he had taken from the table drawer and handed it to me.

Period	BC	Pharaoh
First Pharaoh of Egypt	3138-3087	Naumer
≈		
Great pyramids built	2585-2560	Khufu
≈		
	2130-2120	Akhtoy
	2120-2081	Merykane
Time of the great flood	2081-2075	Mentuhotep I
	2075-2065	Inyotef I
	2065-2016	Inyotef II

White Light Dark Night

	2016-1957	Nebhepetre*
	1957-1945	Mentuhotep II
	1945-1945	Sankhkare*
	1945-1938	Mentuhotep III
	1938-1938	Mentuhotep IV
	1938-1909	Amenemhet
	1909-1875	Senwosret I
Time of Abraham in Syria	1875-1842	Amenemhet II
Noah dies at age of 950	1842-1837	Senwosret II
	1837-1818	Senwosret III
	1818-1772	Amenemhet III
Israelites go to Egypt	1772-1763	Amenemhet IV
	1763-1759	Sobekneferu
Hyksos conquer Egypt	1759-1523	Hyksos kings
Hyksos driven out of Egypt	1523-1503	Thutmosis I
	1503-1492	Thutmosis II
	1492-1471	Hatshepsut*
	1471-1424	Thutmosis III
Birth of Moses per Bible	1424-1381	Thutmosis IV
The Exodus	1381-1370	Amenhotep I
Forty year in desert begins	1370-1361	Amenhotep II
Birth of Moses per history	1361-1352	Amenhotep III
Aten - first monolithic God	1352-1336	Akhenaten
King Tut - the boy king	1336-1332	Tutankhamen
	1332-1313	Ramesses I
Taking the Promised Land	1313-1300	Seri I
	1300-1233	Ramesses II

* Queen during the time her son was too young to rule
Biblical time of the Israelites time in Egypt

"This is the complete chronology of pharaohs who reigned from the time of the great flood to the time of the taking of the Promised Land. The timetable is exact and is taken directly from the hieroglyphs in the tombs. You will note there is no interruption of pharaohs during the time of the flood in which all of humanity, except those aboard the Ark, were said to have perished."

There is only one God

"This table shows a remarkable coincidence concerning a fundamental milestone in the evolution of religion.

"Until Akhenaten's time, all people of the western world, whether they were Egyptian, Greek, Phoenician, Hyksos, Syrian and even those black tribes from the south, worshipped multiple gods. When he came to power in 1352BC, Akhenaten declared, *'There is only one God, the sun God Aten.'*

"Either by coincidence or plagiarism, this became the cornerstone of Moses' religion just a generation later, *'There is but one God.'* The Bible contends Moses lived to the age of 120 at a time the average lifespan was less than 30. This places his biblical birth at about 1400BC in order to allow time for the forty years he was said to have been in Egypt, the Exodus and the forty years in the desert and the time Joshua took to take the Promised Land.

Author's collection

Aten, the Sun God 1352BC

Moses and his God a generation later

"History tells us something much different. It tells us at Moses time, the Israelites were living in Syria, quite a distance from Egypt. It makes historical sense Moses had traveled to Egypt, shortly after Akhenaten's reign, as a young man and learned of Egyptian culture which inspired him to tell his stories, the most fundamental of which he plagiarized from Akhenaten *'There is but one God.'*

"In traveling from Syria to Egypt he would have passed south through the Promised Land. Recognizing its fertile land and its rich abundance of fish of the Mediterranean Sea, he decided at that time to take it for his people.

"In order to convince his people to murder the hundreds of thousands of innocent men, women and children who lived there, he fabricated his many tales, each one designed in progression to substantiate his God's instruction to take the Promised Land, *'Take to the sword all the inhabitants thereof, leave not one alive,'*

"It is that Moses placed himself as the main character in his stories, many of which took place decades before he was born; he claimed to have lived 120 years.

"Nevertheless, history places his lifespan between 50 and 60 years. To think otherwise is to believe Moses actually lived 120 years and that it was mere coincidence he happened to declare *'There is only one God,'* just a single generation after the Pharaoh Akhenaten had come up with the idea.

The resurrection of the dead

"The pharaohs give us another insight into the evolution of religion in the western world. Since 200000BC, all homo sapiens of both hemispheres believed only the spirit could survive beyond the grave. It was obvious to the most meager of dimwits the body would not survive after death.

"Today, all eastern religions, including Hinduism and Buddhism, continue to believe that only the spirit will survive. These ancient eastern religions originated centuries before those in the east first became aware of the western ideology that the body would rise again after death; about a century after Christ's time.

"In 3100BC, the Pharaoh Naumer originated the ideology the body will rise again after death; thus we have preserved for us the Egyptian tombs. This evolved into what is perceived to be the afterlife by all western society; both Christianity and Islamism believe the body will rise again after death.

"This is seen clearly in that all western primitive societies which survived unscathed into the twentieth century, including some African tribes and the Bushman of Australia to the south and the Eskimos to the north, believed only in survival of the sprit. That is, until they became aware of Christianity and Islamism."

"Why do you say, since 200000BC?" I queried.

"Religion evolves like any other social practice. We are the only animal species that buries its dead. The practice dates back to the Neanderthals. They buried their dead, not for purposes of hygiene, but rather out of fear of surviving spirits; the earliest belief in an afterlife. The bigger the man, the deeper he was buried, up to depths of ten feet, to protect the living from his spirit; the assumption being the bigger the man the more dangerous the spirit. In those days, body size meant most everything.

"The concept that the spirit could survive beyond death originated around 200000BC. Archeology has found that earlier Neanderthals and their predecessors did not bury their dead.

"This, of course, tells us the Neanderthals, though not our genetic ancestors, were human beings. As I said, no other animal specie buries its dead. This includes the surviving higher primates - the chimpanzee, the gorilla and so forth."

I surmised, "So belief in God dates back a quarter-million years, interesting."

He corrected me, "Belief in the spirit world dates back that far. Not belief in God. The earliest belief in a Supreme Being, archeology has yet been able to determine, dates to about 75000BC; altars of bear skulls in Southern France suggest caveman worship of a bear god. I might point out, in addition to burying their own kind, some of our early ancestors buried animals, only one kind of animal, bears. They believed bears, too, had spirits which could survive beyond death.

"The greatest war ever fought by human beings did not involve other human beings. It spanned a two hundred thousand year struggle of man and bear for the caves which were their common dwellings. Nevertheless, by the time Moses came along, altars of worship had already become a way of life, as they are today."

Carbon dating analysis

I decided to correct him concerning the dates in the tombs. "These dates of the pharaohs are not etched in stone, you know. I understand the carbon dating analysis method used to determine ancient dates has a significant margin of error." He looked at me as if I had not yet finished grade school

"You think the dating of tombstones is a twentieth century practice? The Egyptians developed the calendar based on a 365-day year in 3312BC. These dates are etched in stone. Perhaps not on every cornerstone, but nevertheless they can be deciphered from the hieroglyphs in the tombs. Carbon dating analysis which is normally used to determine time of a cadaver is rarely necessary in the case of a pharaoh.

"The dating contained in the tombs themselves gives us this information. As a matter-of-fact, it was that the Egyptian tombs were dated as early as 3100BC which confirmed the high degree of accuracy of carbon dating analysis to begin with. Carbon dating analysis has been performed on many of the pharaohs and it has yielded dating consistently within a quarter-century of the actual dates etched in the corresponding tombs.

"We also know the date of birth and date of death and period of reign of those few pharaohs who have not yet been found; this record is in the hieroglyphs of the tombs of their parents and offspring."

I went after him again, "Well, maybe the Egyptians had a law that only Egyptian symbols and writings could be put on tombs? Also, Jews would not have been mummified so today there would be no remains."

"You are right," he agreed, "the bodies might not remain, but the skeletons would surely be there. Under normal conditions it takes a minimum of fifty thousand years for a skeleton to fossilize and even then the fossil would remain. Under desert conditions, the deterioration process is much slower.

"Also, one is not entirely dependent on the markings of a tomb to know its contents. The skull of a Jew is quite different from that of an Arab or an Egyptian or a Greek or a Nubian for that matter. Even an amateur anthropologist can tell one from another at a glance. We know that there are no Jews buried in Egypt.

"One also has the forty year wandering in the Sinai Desert which followed the Exodus. Of course, it is fairly well known the Hebrews were never on the Sinai Peninsula; a number of well publicized explorations before us did that job quite thoroughly.

"Many of these were financed by the Vatican in a desperate attempt to prove some tiny evidence the Israelites had been there. They dug up half of the peninsula in search of some fragment of pottery or some grave marking or some skeleton or anything that would give the slightest evidence that the Israelites were ever there. Again, not a single trace could be found.

"The huge excavation of the Suez Canal failed to yield as much as a finger bone despite the fact that most of the million or so Hebrews who were said to have crossed the Red Sea died during the forty-year wandering; the average life expectancy at the time was less than thirty." He paused for a moment as if reaching for a thought and then went on.

"Actually, I had become somewhat aware of this while I was still in the seminary. A group of us had taken a pilgrimage to the Holy Land to trace our heritage. One day in a coffeehouse, I asked our Jewish host, a professor at the Hebrew University, if there were tours into Egypt that afforded the Jews the opportunity to trace their heritage. After all, the four hundred thirty years they spent in Egypt marked the birth of the Hebrew nation.

"He laughed and told me, 'We Jews believe the Old Testament to be mostly just a story. Yes, important to our heritage and our way of life, but it is simply folklore. Our origins were in what was known then and still is now the *Fertile Crescent*, that part of the Middle East which is today occupied by lower Syria and eastern Turkey; a land of plenty. Egypt on the other hand, has had a history of perpetual famine and starvation. That's why they built the Aswan High Dam in 1971, to bring an end to what had been an eight thousand year history of frequent starvation. It would have made no sense for the Israelites to have abandoned the most fertile land in the Middle East, the land of Abraham, and travel five hundred miles south to live in a desert.

"'Except for the limited area that borders the Nile, which was available for cultivation and harvesting only four months of the year, Egypt was then and still is a desert you know. Egypt's fertile land was flooded eight months of the year and this is one reason why the great pyramids were built as the people had little to do from July to February. It would have made no sense for the Egyptians to let the Israelites in, as they had limited fertile land.'

"'Also, the crocodile and hippo population of the Nile left very few fish available for human consumption. It would make no sense at all for the Jews to leave the Mediterranean Sea and the great

Euphrates River with their abundance of sea life and go to live in the desert.'

"He told me in the Holy Land itself there are many stone etchings dating back to Moses time. 'But you won't find a single etching that depicts the Exodus or the Red Sea miracle of any of the stories Moses told. Particularly surprising is that there are no etchings of the Ten Commandments, not even of the Ark of the Covenant which allegedly disappeared somewhere along the way.'

"He added, 'Actually, any man with half a brain would know that the Israelites could not have lived side by side with the Egyptians in the same land.'

"'Why do you say that?' I asked, puzzled by the thought.

"He was quick to answer my question, 'The Egyptians believed Pharaoh to be in direct intercession with Aten, the Sun God; the reason why the Egyptians lived in servitude to Pharaoh. This is why the great pyramids were built and we know the last of the great pyramids and the greatest temples were built during the time that the Bible claims the Israelites were in Egypt. Yet, there is no mention in the Old Testament the Israelites built any of the temples or pyramids other than that during the last forty years when they built the cities of Pithom and Ramesses under bondage.'

"'The Hebrews, as you know, did not believe in the Sun God or for the matter in Pharaoh's intercession with the Sun God. They, like we today, believed in Moses and Abraham and their intercession with God the Father. To conclude that more than half of the population of Egypt lived in freedom for four hundred years, while the Egyptians themselves lived in servitude to Pharaoh and sacrificed their lives to build the pyramids and the great temples of Egypt makes no sense.'"

Jack went on, this time as the proverbial snowball rolling down the hill, "We now know, as an absolute matter-of-fact, the story of the Israelites in Egypt was a fib. This, if nothing else, tells us the story of Joseph and his brothers, and the story of Moses being found floating in a basket in the Nile, and the story of the Ten Plagues, and the story of the Exodus and the Red Sea miracle, and the story of the Ten Commandments on Mount Sinai, and the story of God's instruction to Moses to take the Promised Land from the Canaanites, and all the rest of Moses' tales were simply stories; outright lies. Of course, this confirms the findings of astronomy, archeology, history and genetics, which today have proved that the stories of Creation and Adam and Eve were just folklore. For these stories too were told while Moses was said to have been in Egypt.

"Without the benefit of what I have told you, it is common sense today that the Israelites were never in Egypt. For had the Israelite nation originated in Egypt, as the Bible contends, the Jews would be making pilgrimages there to visit the gravesites and other Jewish history which is there; to read the hieroglyphs which detail the way of life of the Israelites during the four hundred thirty years Moses claimed they were in Egypt. But there are no Jewish pilgrimages to Egypt, because there is nothing there!"

Since then, I, too, have visited Egypt many times. I, too, have visited the great cemeteries that are located in the cities where the Bible claims that the Israelites lived. I too, have found nothing, not a single trace of Hebrew history there. You, too, will find nothing, for they were never in Egypt. Go there sometime if you are a poor believer. Rather go there if you are a good believer. Go there and search out your heritage as a Christian and you will not find it there.

Actually, to know this today one would not have to go to Egypt because the television history channels have capitalized on the many Egyptian tombs and temples. They take the viewer into hundreds of these tombs and temples and never is there a reference to the Israelites who are said to have lived there; not a single trace of the Hebrews which the Bible contends comprised more than half of the population of Egypt for four hundred years.

It is for certain, if there was the slightest trace of evidence, these entrepreneurs would capitalize on it. After all, their audience is a Judea-Christian world. But they never mention the Israelites, because they know the Jews were never in Egypt.

Yet, the typical Christian still believes the first five stories of this great building of faith are there, because the low lying 'buildings' that surround it, his belief, will not let him know that in fact they are not there. As a matter-of-fact, this great building of faith has not a single truth to stand on. If one is searching for the truth, he or she should not be listening to what someone of motive has to say, as is the case with Moses, but one should embark on a trip to Egypt where lies, the truth.

Customers who consider paying a preacher who peddles the *God of Moses* as the true God should ask themselves, "If this preacher insists the Israelites were in Egypt as his Bible claims, why should I believe anything else he has to say?"

I sat there for some time thinking of what this meant as if I could figure out some way around it. "What this means, what this means," I spoke his conclusion in a series of definitive statements, "is the Old Testament is simply folklore and is not the word of God, that as a matter-of-fact it was not even inspired by God, that Jesus Christ claimed to be the *Son of God* who was sent to earth to rid the world of *Original Sin* as was once foretold in a fairytale!"

I peered around the pond. It was quite apparent to me that every animal, bird and even insect had come to a screeching halt at what I had just said. Even the fish in the pond had slowed up a bit. I thought at any moment they would attack and rid the world of the infidel who sat before me and I feared they might take care of the witness at the same time. Yet they did nothing; just remained there in a state of shock with frozen stares.

"They must all be on his side," I thought.

Then Jack wrapped up his conclusion, "The stories of the *Israelite's Enslavement in Egypt,* of the *Exodus,* of the *Red Sea Miracle*, of the *Ten Commandments*, of *Noah's Ark*, of *Creation,* and of *Adam and Eve* are for mothers and fathers who want to bring up their children in a world of make-believe. They prefer to play with the minds of children beyond the age of six as their parents have done before them. They bring them up in a world of mysticism rather than in a world of reality where they could better develop their minds and make their contribution to society."

He sat pensively for a few more minutes, anticipating any question I might have. I peered around the nature that was all about me once more and wondered if there could be a spy or two among the birds or the bees. Finally, lowering my voice to a whisper, "Who knows the results of your investigation?"

He replied, "Only a small handful of cardinals clustered in the Vatican; the Pope, the Secretary of State, the Undersecretary of State, the Foreign Minister, the Deputy Minister, the Primate of Opus Dei and, of course, the three of us who conducted the survey. That is, two of the three of us who conducted the expedition."

"Why do you say two of the three of you? All three of you must know. After all, you were all there," I corrected him.

Instead of answering my question, he looked me square in the eyes, "This is where you come in."

"Where I come in?" I questioned.

"Yes, you," he replied. "Now, you also know." I kind of shuddered when he told me. I again peered around the pond. Jack sensing my diversion remained silent as to permit me to check out each of the creatures that might be listening. He started up again.

"Piccolo's intention is to bring an end to this kind of injustice once and for all. To take the voice out of the Bible that tells one to hate one's fellowman. He intends to take from the *Fascist* preacher, as Piccolo puts it, *'his flimsy arrows built in an ancient factory.'* He has now struck on the strategy of battle. He intends to pull the rug out from under Moses; to annihilate his great arsenal of hate. But in pulling the rug out from under Moses, one must take care not to also pull the rug out from under Christ.

"Now, we have finally arrived at the point at which I can tell you of the seriousness of the problem we face." He paused for a moment as if collecting his thoughts before going on. He glanced around as to be certain that he could trust all of nature's creatures that had gathered around him, as if to assure himself there was not a spy among them.

A body floating in the canal

"A few months ago Brother Tom Jones was found floating in the canal that runs beside the cathedral. His real name was not Tom Jones. I can't even begin to pronounce his name. When inducted into the order, he had picked a long entangled name of an ancient philosopher; because it was tough to remember and almost impossible to pronounce everyone called him Tom Jones.

"The official pronouncement was he had drowned; despite the fact he was known to have been a good swimmer. Piccolo grew suspicious and had an autopsy performed. As no water was found in the lungs, the results of the autopsy proved he was already dead when he went into the water. There had been hemorrhaging in the brain; he had been clubbed with a blunt instrument. Piccolo's suspicions were well founded, Tom Jones had been murdered.

The canal that runs along the church where Tom Jones was found

"Piccolo told me that we ourselves could be in danger. He was particularly concerned about me, as being a member of the council and a bishop of the Church he was far less a risk to the conservative core in the Vatican than was a rank and file member of the clergy who was more likely to stray to the press. After all, if Piccolo, a potential cardinal, were to go to the press, it would destroy his own ability to eventually accomplish the things he wants to accomplish to make this a better world to live in.

"Piccolo told me he had called the Pope and told him of the autopsy findings and Paul had instructed him to relocate me from the seminary dormitory here into the bishop's castle. The house has several bedrooms which are normally reserved for visiting prelates. At the time I moved in, Piccolo had a half dozen unwed mothers living in the house and because of the impending danger he relocated them to adjoining buildings. The Pope told the members of the *Vatican Curia* of the autopsy findings and warned if there were any more mysterious shortcomings in the Veneto country, he would order a full scale investigation.

"Paul's action gives us some protection as an investigation could lead to disclosure of the findings of the Egyptian survey. It would defeat the purpose of any further murders since such a revelation would bring about the destruction of the twenty or so cardinals that make up the *Vatican Curia*. Of course, if something were to happen to Paul, we would surely be in great danger.

"On Piccolo's directive I have told you of the results of our investigation, results that are known to only a small handful of men, results which could imperil Christianity should they become known prematurely before the Church is able to complete the separation of Christ from Moses. I have told you of our own possible demise. I will leave that much with you for now."

I quickly corrected him, "To separate Christ from Moses is not possible. Christ claimed to be the Son of the *God of Moses*.

"I will answer that one in a moment," he decided to close the lid on Moses. "Today, many Christians who want to seal their faith make a pilgrimage to the Holy Land during their lifetime. Luckily they don't go a step further and make a pilgrimage to Egypt, which is just south of the border, for it would certainly seal their fate. If there were any Hebrew tombs in all of Egypt, you can bet the churches and the synagogues would be capitalizing on them, would be making use of them to prove their point. But you never hear of them. The reason you never hear of them is because they are not there." Raising his voice, he pounded the final nail into the coffin, "The Israelites were never in Egypt."[1]

---

[1] Bibles claim between six hundred thousand and two million crossed the Red Sea.

Today, most Christian churches, including the Vatican, view most of Moses' stories as mythology (fairytales). This would include Moses' stories of Creation, Adam and Eve and Original Sin, the four hundred years the Israelites were in Egypt and the Exodus and the taking of the Commandments on Mount Sinai. In many catechisms today the Catholic Church footnotes these stories – Creation, Adam and Eve, Exodus, etc. - with the words: "not a matter of history." In order to preserve some possibility that God actually talked to Moses, the last apparition of God to Moses was the taking of the Ten Commandments on Mount Sinai on the Sinai Peninsular. Late in the twentieth century, when hundreds of archeological expeditions by the Vatican and other churches confirmed the Israelites had never been in Egypt and the 40 year wandering never took place on the Sinai Peninsular, dozens of peaks in eastern Turkey and elsewhere in the Mid East were renamed 'Mount Sinai' – various claims arose that the taking of the commandments took place on these various mountains – the official position of the Catholic Church today. Recent television documentaries claim only a few dozen were involved in the Exodus and the 40 year wandering in order to get around the lack of archeological evidence. Jack's declaration "The Israelites were never in Egypt." would not be as startling in today's world where most people and most Christian churches know the Israelites were never in Egypt.

Chapter 19

The God Zagreus-Dionysus

Pausing for a moment or so as to emphasize his last point, Jack finished, "The survival of Christianity, or for that matter any religion, depends solely on the population's tendency to believe. In the case of Christianity that belief is now in great jeopardy. We must begin now to separate Christ from Moses, to separate Christ from bigotry for all time.

"This is where you come in." He started to set the stage for his conclusion, "If Piccolo was to announce his findings today, it would annihilate Moses and the bigot but it would also annihilate Christ for we have not yet entirely separated Christ from *Original Sin* and Moses. This is a delicate process and one that will take a few more years.

"Piccolo feels in a few decades the job would have been largely accomplished and people will believe in the man in the robes as Christ is portrayed on the motion picture screen and won't care much about His connection to *Original Sin*.

"The theological link from the gospels backward to the *God of Moses* is the absolution of *Original Sin* through the ritual of baptism; the fundamental ritual of Christian doctrine for salvation. Take it away and Christ stands alone free of the evils of Moses."

"The theological link from the Old Testament forward to Christ is the *God of Isaiah* who foretells His coming and promises us eternal life. The *God of Moses* never thought of Christ or Baptism. He left us without eternal life, *Genesis 3 'In the sweaty of thy face shalt thou eat bread till thou returneth unto the ground; fro out of it wast thou taken; fro dust thou art; and unto dust thou shalt return.'*

"One could ask why the evangelists linked Christ to Moses when practically everything Christ had to say contradicted what Moses had to say. The answer is clear. They would have never gotten the business of Christianity off the ground to begin with, because at the time everyone believed in the *God of Moses*. In fact, they were still stoning unwed mothers to death and sacrificing the blood of innocent animals to the *God of Moses*. As a matter-of-fact, the Jews and Muslims still do today. The reason they will go on killing each other to the end of time.

"Hopefully, by the end of the century, the transformation will have been completed and most people will think Christ, like the Greek God Zagreus before Him, *'Is the Son of God who came to earth to die for the sins of mankind and grant us eternal life.'* This will change His link from the *God of Moses* to the *God of Isaiah* which is the explicit testimony of the Bible.

"In ancient Greek mythology the story of the God Zagreus-Dionysus dates back a thousand years before Moses. In the legend, Zeus, King of Gods, speaks, *"Hail thy offspring who come upon thou without lust ...Hail thou who has been counted among thieves and murderers... suffered the suffering ...Thou wilt become God from man and rule eternal life."*

"Zeus wills to send His Son to save the world. The God Zagreus, Son the God Zeus is born of a virgin, a fundamental requirement of divinity. When sent to earth, Zagreus is lured into captivity by the Titans who tear him apart and eat his flesh.

"When Zagreus dies he gives birth to his reincarnation, Dionysus who rules the universe as Christ will come to rule the universe in Christian theology.

Author's collection   Alexandria Library

The death of Zagreus   The birth of Dionysus

"It follows we have the only prophesy in the Old Testament of the coming of Christ, two thousand years after the time of Zagreus, The *God of Isaiah* speaks in *Isaiah 7*. *'Behold, a virgin shall* conceive and bear a son, and shall call his name Immanuel,' followed by *Isaiah 53*. *'He hath poured out his soul unto death; and he was numbered with the transgressors; and he bare the sin of many, and made intercession for the transgressors.'* [1]

"In time, the population will come to accept the natural progression of the *God of Isaiah* to Christ which is explicit in the Bible: Christ like the God Zagreus before Him *'was sent to earth to die for the sins of mankind and bring us eternal life.'*

"Yet, in truth, Christ never died for the sins of mankind, for had He done so mankind would not be at risk to hell. The people in Zagreus' time had never thought of hell as a possibility. That came along later in Moses time. So it made sense in those days *'The God Zagreus was the Son of God who came to earth to die for the sins of mankind.'* But it makes no sense today. We are going to suffer for our own sins and no one else is going to die for them. Also, keep in mind that Christ never really died on the cross."

I gave him a look of apprehension and disbelief.

"Yes," he answered my expression, "the true trauma of death is the great emotional pain of death and not the physical pain of death. Even in the case of physical pain, Christ suffered only a few hours whereas most of us will suffer for weeks or months before death comes as a friend and takes us from our misery.

"It is the realization we are about to leave our loved ones and not knowing we will ever see them again which is the true suffering of death. Also, in the case of the Christian one knows as a matter of faith that one is never going to see their loved ones again, the reason why all Christian marriages end in the phrase *'Until death do us part'* or its equivalent.

"The God who spoke to the prophet Jeremiah, could not have been more explicit, *Jeremiah 16 'Thou shalt not take thee a wife, nor a sibling, nor a son, nor daughter into this place.'*

"But in Christ's case, He, being God, knew for Him death was not the end. Christ never suffered the great emotional pain of death. Rather, for Him, His death on the cross was a time of great joy for He knew that He was about to join His Father in Heaven that very day. For Christ, dying on the cross was a charade. What's more, He knew it.

"Nevertheless, believers never stop to analyze theology, so this is the course we will take. People will no longer care whether or not the story of *Adam and Eve* is true or not. *Original Sin* will be history." He thought a moment and then corrected himself, "No, not history, a fairytale someone once told." An awkward moment followed as I tried to rationalize all that Jack had said in an attempt to somehow salvage my faith. Then he started up again.

"Baptism, which rids one of *Original Sin*, has nothing to do with our own will or acceptance. It most often involves infants who have no idea of what it means. It is a theological fact, no matter what one desires, one cannot baptize oneself. If someone escapes childhood without being baptized, one is still dependent upon another person to perform this ancient ritual. It has nothing to do with what is right or wrong. Believe me, regardless of what scripture might claim, it has nothing to do with salvation.

"No God, in His or Her right mind, especially Jesus, would bar one from the Kingdom of Heaven because someone did not splash water over his or her head and utter a single sentence. Only a fool would believe that. Baptism was put into the gospels in order to cut out the competition. Its motive was clearly to destroy the salvation of other religions and as I said to fabricate a link from Christ to Moses. Always keep in mind that religion is a business and like any other business one has to play the game."

A sacred commission

He paused briefly and started up again, "Nevertheless, in case something was to happen to us, Piccolo feels someone else should know; someone not geographically near us. Someone we could trust. Someone who we could be reasonably certain would be around at the end of the century after the transformation of Christ's purpose is complete; someone who, at that time, could announce the results of the Luciani expedition into Egypt and bring an end to Moses influence in the western world once and for all. Someone who will bring an end to bigotry as quickly as one turns off a water faucet.

"That Piccolo, perhaps long after he is gone, would have carried out his beloved father's sacred commission, *'You must play your cards carefully and work hard until that day when at the helm of their ranks you will establish the common dignity of all God's children in the world. You must find the strength along the way to do what you have to do to bring about a day when all men accept each other as being equal.'"*

I stopped him, "But me? How could I possibly accomplish such a thing? I will probably be a 'nobody' at that time."

He quickly shot back, "You will do it in the very same way every other 'nobody' has accomplished his or her objectives. The very same way Paul and the others accomplished their objectives."

"Paul and the others?" I questioned.

"Yes, like Paul and his contemporaries Matthew, Mark, Luke and John wrote their book many years after Christ died," he paused and then raising his voice, "You, too, will write a book."

"But I don't know how to write a book," I told him.

He shot back, "You will write a book," he raised his voice to the point of a demand, "for each and every one of them!"

My mind drifted back to yesterday, to that time I had taken the pledge in the great cemetery of Milan, "*. . . my solemn promise to carry out that sacred pledge for each and every one of them. That what they dreamed of, those things they willed to be, will come to be, for each of them, and for me, and for you, and for all humanity!*"

The factory

Getting up, he patted me on the shoulder, "Enough said. Let's go to lunch. Vittorio Veneto is smack in the middle of Italian wine country. I will take you to a little sidewalk cafe where the food is terrible, but what counts most, the wine is marvelous."

We returned to the main road. Instead of going up the mountain toward the castle, we moseyed our way down into the village. Passing the cathedral we began a long walk in front of the seminary complex. As we walked along the yellowing nineteenth century stucco buildings he told me, "This is our crown jewel. The seminary is what makes Vittorio Veneto such an important post. It wasn't always that way. Piccolo made it that way.

"When he came here ten years ago it was just another priest factory. Today, although it is exceeded in size by the Gregorian University of Theology in Rome and a seminary in Milan and another in France, it is the most recognized in the Catholic world.

All the others put out good products, anything from Volkswagens to BMWs; we put out a consistent line of Mercedes.

"Piccolo recognizes priests are salesmen of the faith and he believes they are a chain. He believes in the age-old adage that a chain is as strong as its weakest link. If one priest screws up, it will do irreparable damage to all the others. So when he came here, he changed the manufacturing philosophy from one of nurturing the quickest and brightest students to one of concentrating on improving those at the bottom, those who were the slowest and those having the greatest difficulties. Today, each one of them leaves as a qualified *550SEL*, fully equipped to meet anything they encounter on the highway of faith ahead; able to negotiate the most challenging hairpin curves of doubt that they might come upon.

"A very big part of a seminarian's education here is the human mind. This is the reason why I went to med school. Unlike most bishops who surround themselves only with Bible thumpers, Piccolo built the faculty here into a balanced staff of scientists, historians and most importantly, psychologists; and in my case, a psychiatrist. He believes that the main purpose of the priesthood is to help people lead more healthy lives - healthy mental lives. He is less concerned with the afterlife than he is with this life.

"Most people think they go to church to adore God. Yet, in fact, they usually go for other reasons: they think it is a mortal sin not to go to church, they go as a matter of habit, but mostly they go for therapy. Some of them find redemption in the ranting and raving of evangelist preachers and others find solace in the most soft spoken of priests. But, nevertheless, this is what they get out of it, a once-a-week rubdown of the mind. This is consistent with the gospels. Christ tells us many times in His ministry that going to church has absolutely nothing to do with salvation.

"Piccolo found it to be irreprehensible that a psychology course was not a requirement in seminaries; that it was rarely offered as an elective. He once told me, 'The state is a coward when its adversary is the Church. It requires all practitioners of the human mind including psychologists and psychiatrists, to have accomplished a prescribed curriculum usually to the doctoral level before authorizing them to deal with the human mind. Yet, it looks the other way when men disguise themselves as representatives of God. The ministry is nothing more than an army of men totally unschooled in the way the human mind works that spends most of its time dealing with the minds of men, mostly with the minds of weak men. No wonder the world is so confused.'

"Piccolo has urged Pope Paul to do the very same thing that he has done here at Vittorio Veneto. To trade in the *Vatican Curia* - a group of mentally deranged theologians - for a team of experts on people problems, so that the Church can better understand its congregation's needs."

We headed down a cobblestone street and turned right at the town clock. Passing one of those Italian ice cream stands one sees all over Europe, Jack said, "There are over five million of these stands in the world; this one, you will find, is in first place. We will stop here on the way back."

"Not many people," I thought. Yet each one, some walking, some running, and a few on bicycles, had obviously been there all their lives; a hundred, perhaps two hundred Italians and I.

As people looked at me, I relished my individuality. Every one of them, every single one of them, said "Hi Jack" in perfect English despite the fact Jack could speak perfect Italian. Each of them had an expression of great respect and admiration and awe, as if Jack were some kind of a God.

The town clock at Vittorio Veneto

¹ pg 237 For more on evolution of world religions see Lucien Gregoire's *A God for Lions*

Chapter 20

Albino Luciani and General Patton

At one end of the street, where it formed a wedge with its neighbor, we took a table outside a village restaurant; one of those they try to duplicate in the big cities with little success. Heeding Jack's advice, I decided on bread cheese and fruit. He doubled the order and uttered something to the waiter in Italian.

In no time at all he returned. He spread about the table an assortment of breads and cheeses and fresh fruits. He placed in the center of the table a crystal clear bottle of wine having a slight amber touch to it. It was set in chopped ice held in by a clear glass container. I was about to get a taste of the afterlife, itself.

Reaching for the wine, Jack poured our glasses properly half full of this memory which is still with me today. Like any other close friends who hadn't seen each other for a number of years we started to chat away the hour about the good old times.

Somewhere along the way I asked him, "Boy, you sure are popular. Just being the bishop's secretary gets you that?"

"It has nothing to do with that," he answered. "It is due to something else entirely."

A thousand 'hellos'

"My uncle was in the military during the world war. He held two purple hearts and a dozen other enviable decorations. In a heroic action he was hit in North Africa and was hospitalized for six months with serious liver damage. He was nominated for the *Medal of Honor*, but instead received a *Silver Star*.

"Anyway," he continued, "a year later he was hit again and this time he was paralyzed from the waist down. He was hit by friendly fire when he placed himself between his own troops and twenty-eight Italian school children so as to alert his men they were about to fire on the children. It happened outside Milan quite a distance from here

"He was again nominated for the nation's highest award. Because it had occurred toward the war's end, it helped to heal the wounds of war and since his action had been witnessed by so many, we were certain this time he would reap the honor that had escaped him in North Africa.

"Wow. He won the *Medal of Honor!*" I exclaimed

"Not quite," he corrected me. "This time they gave him the *Distinguish Service Cross*, the nation's second highest award."

"But, I don't under . . ."

He cut me off, "The *Distinguished Service Cross* is for extraordinary heroism involving risk of life 'not quite' justifying the *Medal of Honor*. The 'not quite' in my uncle's case was determined the same way as it was in the case of many other candidates who were passed up for the highest award."

"How's that?" I queried.

"During the world war, 'not quite' was often defined as being soldiers of color and of certain ethnic backgrounds. In my uncle's case 'not quite' meant something else."

"Something else?" I repeated.

"Yes, something else. My uncle was gay and because his life then seemed wasted he felt he could be a martyr for the gay cause. Foolishly, he announced he was gay while the matter was still pending. He once told my mother his admission of his homosexuality took much more courage than when he had placed himself before the firing squad to protect the children. He hoped he could attract national attention and do something for the gay cause, but he didn't attract a fly. All he accomplished was to deprive himself of the nation's highest award.

"Nevertheless, this is why I am a celebrity in these parts. This is why wherever I go I get a thousand 'hellos'. The people believe some of my uncle's courage has rubbed off on me.

"Back then, homosexuality was not the type of thing his family wanted to talk about, much less see spread across the front page of the local newspaper. In fact, it still isn't today."

"So, the Army threw him out?" I queried

"No, up until the second world war, there was no policy excluding homosexuals from serving in wars all the way back to Washington's time. Through the years, many known homosexual men and women enlisted and served honorably. There was no box to check, no policy; and this continued into World War II.

"When you have a real war, one in which you are fighting for survival, one doesn't stop to sort out what the preacher considers to be morality. After the war, things began to change.

"That he didn't attract a fly was not entirely true. He did attract Eisenhower's attention. He drew an official reprimand directly from Ike; something General George Patton, my uncle's commanding officer, tried to block.

"Actually," he continued, "my uncle relished in it. He saw it as a victory; that his action drew the attention of Ike, himself."

"So how is he now? Certainly he could do something now; being a war hero," I offered.

"He isn't," he replied, "A couple of months after receiving the award, he died quite suddenly from a liver infection related to his earlier injury that was brought on by his confinement to bed.

"Freedom without equality is not what it pretends to be . . ."

"Only his father flew to Italy for the funeral. His mother was too distraught. His father told my mother the army provided full military honors; there was some speculation General Patton himself, who happened to be in the area at the time, would attend. Although several hundred villagers attended the ceremony, he didn't show up. In fact, not a single officer of rank showed up, despite the fact my uncle held one of the nation's highest awards. Yet, someone else did show up."

"Someone else?" I queried.

"Yes, Piccolo showed up. It is not by mere chance I managed to secure an assignment here as an instructor in this remote Italian seminary and eventually grew so close to my patron. Piccolo had officiated at my uncle's interment. Yes, I groomed myself for the assignment, going so far as to earn a PhD in psychiatry and becoming fluent in Italian, but the only reason I got the job was because Piccolo had rendered my uncle's eulogy.

"At the time, Milan was a conservative diocese and no local priest could be found who was comfortable in officiating at the ceremony. The archbishop reached into the Veneto country which was, as it is today, the most liberal part of Italy. The choice was the revolutionary and outspoken young priest from Belluno.

"As I said, the army provided full military honors for my uncle. A lance corporal showed up heading up a detail of four soldiers, three with rifles and one with a bugle. Piccolo doesn't remember too much of what he said, other than he had ended his string of prayers with an embittered comment, *'It is the soldier who shed his blood on the field of battle and not the preacher who cowers in his pulpit, who should determine who should or should not be free.'* [1]

"The detail raised their rifles and fired off three volleys and the man with the bugle did the best he could with the taps.

"Three volleys for a man who gave his life to save others. Yet, congressional rules call for nineteen volleys for the political appointees of a president. It kind of tells one how America thinks.

"Anyway, as the crowd of several hundred villagers started to move away from the grave, a strange thing happened."

"A strange thing?" I repeated his statement as a question.

"Yes, a large olive green military sedan, one of those with the large bubble fenders one often sees in the movies, entered the cemetery and moved along its outer perimeter and finally came to a stop in the roadside that annexed my uncle's grave. A young soldier got out of the front seat and opened the rear door and an army officer of about sixty stepped out.

"When Patton reached the gravesite he introduced himself to Piccolo and my uncle's father as if the rows of stars on his shoulders could not have conveyed the message. A light rain was falling from darkened skies and one could imagine hearing the firing of artillery shells in the distance although there was nothing there. Today, a quarter of century later, Piccolo remembers every single thing the general had to say.

"The great general told them, speaking decisively and pausing on every syllable, as if he were addressing Congress, 'At West Point, there are many courses, one learns many things: one learns the history of war, one learns the purpose of war, one learns the strategy of war, one learns the struggle of war, one learns the noise of war, one learns the horror of war, one learns the victory of war, one learns the hopelessness of war. Yet the most important course one takes is taken on the great battlefield itself. That course is called *courage*.

"He continued, 'As you know, there is a great difference between Ike and myself for Ike has never taken this course. Never once in his lifetime has he carried himself into battle, into the pit. Not once in his lifetime has he pulled the boy out of the mud and searched for where the mud left off and the blood began. Not once has he reached for the final pulse of this thing called life. Never, not once, has he given himself the great opportunity to realize this great prize we know as courage. His only experience in battle has been in his reading and in his textbooks and in his toy soldiers and in his toy tanks and in his toy ships that he moves about on his great table of war. Like the preacher in the pulpit whose only time in battle has been in the atrocities of the ethnic cleansing wars in his scripture.'

"With a great tear forming in the corner of his eye, Patton placed his hand on my uncle's father's shoulder and told him, 'I apologize to you for Ike's action. His action does not speak for me, does not speak for those who fought along side your son, does not speak for America. It certainly does not speak for freedom.

"'Your son,' he went on, 'has won for you and all of us, this thing called courage. This thing called courage that I, too, have sought many times. That I, as a commanding general, would place myself upon a tank at the very forefront of my ranks in open line of enemy fire and with artillery shells bursting all around me, I too, have craved for the taste of this thing called courage. Yet, even I have yet to realize its dream.

"'Yesterday, I assigned my highest decorated and most courageous officer to represent your son's fellow soldiers and America here today. Late this afternoon, I learned that officer had failed in his duty by delegating his sacred responsibility to a subordinate. Not because he had more important things to do, but because he didn't have the courage to be here. A different kind of courage, a greater kind of courage; he was afraid of what the press might do to him, of what his fellow soldiers might think of him, of what his family might think of him, of what America might think of him.

"'Your son had that kind of courage. He wasn't afraid of what the press might do to him. He wasn't afraid of what the army might do to him. He wasn't afraid of what his family might think of him. He didn't even care what America might think of him.'

"The general then stepped a few paces to the left and placed his hand on the shoulder of the adjoining tombstone, the one marked

Anthony Jackson, Bronze Star and continued, 'Even today the *Christian right* continues to persecute southern blacks, many of whom have also shed their blood on the great battlefields of this war. It continues to practice its bigotry against Jews and women and homosexuals and many others; all those who appear to be different.'

"He then placed his hand on the shoulder of the next stone, the one marked *Patricia Wilde, Bronze Star*. 'The world will never know of her valor, never know of her bravery, never know of her courage, never know of her gallantry; never know of these things that had won for it its freedom. All that evidences she had ever been, is this small white granite stone and the marks upon it.'

"The general then turned toward the crowd and raised his voice, *'After this war is won, after the final volleys are fired, after the smoke clears and the tears begin, America must fight a new kind of war and that war will be fired by a new kind of courage.*

"*'This war will win for America and all mankind this thing called freedom. But that war, the war within, will someday win for*

America the great prize of equality for all men and women, something that this war cannot do. For freedom without equality is not what it pretends to be. The diamond would be made of paste.

"*'It is our differences that have made us into the great nation of one that we are and the day will come when men and women of good conscience will no longer heed the words of those preachers and politicians who choose to use them to divide us.*

"*'When that day comes, all children will be born into a world of equal opportunity. Only then will America and the world be truly free. Yes, today men and women of great courage are engaged in this great war which will soon crush the enemy from without, but it must take many more men and women of still another kind of courage, a greater kind of courage who must rise up and crush the enemy from within. Only then would these brave men and women not have died in vain. Only then will the diamond, this thing one calls freedom, be real.'* [2]

"As he climbed into the vehicle, Patton told my uncle's father, 'I have been proud to have had your son serve in my army here,' looking upwards toward the sky, 'and I am certain He is proud to have him in His army, today.' Letting a tear drop onto his cheek, he nodded to his driver, and they were off."

Jack went on, "It is the great regret of my family, that in order to save themselves embarrassment, they intentionally kept the situation quiet; they intentionally kept it from the press. They intentionally kept my uncle from making his mark. His Mom and Dad still live with this, their inaction, which they now know was an atrocity on their part. But there is now no remedy. It is too late. Their lives are now broken. They have aged two lifetimes in one. Even I, though only a child at the time, have great regrets about it.

"But, you're right. If he were alive today, he could do something; perhaps not much, but something. In fact, all the other homosexuals who hold high military awards including many who hold the *Medal of Honor* could do something."

I interrupted him, "I thought you said they weeded out the gays in awarding the *Medal of Honor?*"

"The great majority of gays are still in the closet today, and even more so was the case then. In the war, over four hundred congressional medals were awarded. If one plays the percentages, this would mean about thirty to forty of them are held by truly homosexual men. But of course, like my uncle most of them did not survive. But if one or two of them who did survive were brave enough to come forward, it would stop much of the preacher's bigotry in its tracks. But, as my uncle said, that would take more bravery today than did his action in 1944 when he stood before the children.

"This is what Patton was referring to when he said *'It must take many more men and women of still another kind of courage, perhaps even a greater kind of courage, who must rise up and crush the enemy from within.'* Celebrities who make public their sexual orientation serve as great role models for gay youth. Bring it out in the open and the stigma will disappear overnight.

"Actually, today in Vietnam, there are a disproportional number of gays versus straights serving and giving their lives for this thing we call freedom, as the current draft law exempts men who are fathers of young children. You've done your time in the military. Did you ever meet a WAC who was not a dyke?"

With a gesture of reverent solemnity, he stretched out his arm and pulled back his sleeve to display his wristwatch. On one side on the watchband was welded a small heart pendant with an image of George Washington on a purple background. On the opposite side of the face of the watch was a gold star.

"The *Distinguished Service Cross* and all the others are buried with him. They are in the great cemetery at Milan. There is something more.

"Three years ago, Piccolo used his influence with his good friend Aldo Moro who was then Prime Minister. On March 29, 1965, at the gravesite with twenty-three of the surviving children looking on, on behalf of my uncle I accepted the citation from Moro that is on Piccolo's office wall. Moro, quoted Patton in his closing remarks, *"Freedom without equality is not what it pretends to be, the diamond would be made of paste!"* As you know this became the essence of Aldo Moro's struggle."

Noticing the gold star, I said, quite confused, "I thought you said it was a silver star?"

"Look closer," he replied, "there's a tiny silver star in its center." I looked and sure enough it was there.

Suddenly, at the very same instant, a light bulb went on in my mind. "Milan," I exclaimed, "Was his name Champney?" I asked, at the very same instant realizing it couldn't have been, as it was his mother's brother we were talking about.

"No," he answered, "Phillips. Frank Phillips." Judging he had struck a nerve he fell silent. As I stared out onto the street, I recalled the etchings that read: *Frank Phillips, Distinguished Service Cross; Anthony Jackson, Bronze Star; Patricia Wilde, Bronze Star* and the others.

Leaving some of the wine in order that we could find our way, I settled up with the waiter and we started back. Stopping in front of the gelato stand, I followed Jack's lead, "one of these, one of these, one of these and one of those." As we walked toward the seminary, I thought in great wonderment of how the finest gelato in the world could have found its way so near to its finest wine.

As we came to the end of the street he went one way, and I the other. He called out after me, "Remember, dinner is promptly at seven-thirty."

As I walked back to the hotel, it ran over and over again in my mind, *"To carry out that solemn promise, to answer that fervent prayer, to carry it out for each and every one of them, that what*

Lucien Gregoire

they dreamed of, those things they willed to be, will come to be, for each of them, and for me, and for you, and for all humanity!"

Dear Mr. Gregoire,

My name is Tommy. I read Murder in the Vatican..

I was born with one eye and had a wrinkled face so nobody wanted me. I think I scared them. Then one day when I was five years old my father showed up and took me home. I remember the kids used to laugh at me in the playground. Then my fathers sold everything they had including the house and I spent a long time in the hospital and the doctors and nurses made me look good. We have still many bills to pay. But now I am quite a ladies man at school. I hit my first home run last summer and this year I am going to hit forty more.

My parents love me. That is, they used to love me. Now only one of them loves me. Because the other one is dead. He gave his life trying to win freedom for Iraq. He died on my fourteenth birthday. I noticed in your book that General Patton said some words over the grave of another soldier who like my father was gay and gave his life to save some Italian school children, "Freedom without equality is not what it pretends to be. For the diamond would be made of paste."

As you know I won't be getting any benefits from the army to pay for my education or my medical bills because my parents did not have the freedom to marry and I was adopted by the father that is still with me. I didn't even get his purple heart or bronze star. The army gave them to his parents who hated him and sold them in a tag sale. I would have liked to have them. And as you know I am also not entitled to social security and other benefits despite that my father like all other fathers paid for them. In his case not only with his dollars, but with his life.

My father tells me not to feel bitter because according to the last census there are a million children in gay families and they are in the same boat. But, of course, I am not listed with them because of the don't ask don't tell policy. My father says that there are also over fifteen milion other children in single parent families, many like me who are of gay parents who are not counted in the census because they are fearful of losing their jobs too. So we are talking about a lot of children here, million s of them, not just me.

Both my father and I are sure that your book will be a good seller. I want to ask you a favor. I want to ask you when you are on Larry King Live, could you please read my letter. I think it might help people to understand our problem.

I was planning on being a baseball player. But your book has made me change my mind. Instead I am going to study hard and I am going to grow up to be president. Well, maybe not president, but I am going to help other people. Like me.

---

[1] pg 245 Albino Luciani Milan Italy 29 Mar '44 Luciani repeated this statement many times during his ministry

[2] pg 247 & 250 Milan Italy 29 March 44. General George Patton. Albino Luciani, who had officiated at the same ceremony, quoted Patton's remarks on October 17, 1973 in a sermon in the Basilica San Marco in Venice. The unitalicized remarks are to the best of Albino Luciani's recollection as he related them to the author.

Chapter 21

Mud in the Street

I found myself in an ancient room of stone. Its focal point was a row of bottle-bottom leaded glass windows peering out through ornate iron grates. They were flanked on either side by heavy wooden shutters which opened inwardly into the room and were set upon encasements that projected outward from the wall; so much so that if the shutters were closed one would think one was looking at a cabinet. Although their leaded frames obscured the view, it was clear what they were looking at; surviving ruins of the surrounding castle towers.

The old stone flooring with its centuries of wear was every bit as cold as it seemed to be. A beautiful sideboard of mellowed wood that had to be several hundred years old ran along the wall opposite the windows.

Just above it an aging, yet quite colorful, tapestry of princes and paupers blanketed the wall. At the far end of the room a large oil painting in a heavy, gold encrusted frame depicted Christ's miracle of the loaves and fishes. I correctly assumed it was another Titian.

At the other end of the room was a giant walk-in fireplace cut into the stone wall. A giant black iron cauldron hung in front of it exactly as it was on the day it had been forced into retirement by the stove that stood in the kitchen off to one side.

In the center of the room was a carpet of what must have been extraordinary value. On it stood four richly carved chairs that clashed with the table they surrounded. As a matter-of-fact, everything in the room including the room itself clashed with the table. The table was one of those cheap, browning, enameled top tables of the forties. Four chairs, I would guess, would very easily draw twenty or thirty thousand dollars at auction, hemmed in a table that would go for five dollars at a yard sale.

Jack explained, "When I mentioned this afternoon that Piccolo started to sell all this stuff and the Vatican stopped him, I wasn't kidding. Yet, they didn't stop him in time. His first visitor from Rome was Bishop Caprio. When we led him into the dining room and he saw the table was gone he threw a fit. He walked about the room thrusting his arms into the air like a man out of his mind. He ordered Luciani to stop this madness, this illness *'Sell all that thou hast and give to the poor.'* "

The orphanage beneath the bishop's castle

I stopped him, "You mean he built the orphanage at the foot of the mountain with money he got for a table?"

"No," he told me. "He built that one with money that was intended to build a church. A wealthy man who was terminally ill intended to will his fortune to build a church, but Piccolo talked him into building an orphanage instead. Piccolo considers a church a luxury and to a certain extent unnecessary, particularly when Christ's children are starving to death as we speak.

"He once told his congregation here in Vittorio Veneto, 'Nowhere in all of Christ's testimony does Christ ask we fall down on our knees and adore Him. Rather, Christ spent His ministry helping others less fortunate than Himself and asked us only to live in imitation of Him. Those of us that think we can butter Him up by building great edifices to His honor and falling down on our knees and praising Him, as did the Pharisees in the temple He so often condemned, believe me, are fools among fools.'

"As I've told you before, Piccolo's authority comes directly from Christ. *'Now, what would Jesus have done in this case?'* Piccolo told me that as long as a single child lacks a roof over his or her head and has not enough to eat, not a single block of stone will be laid in Italy to build a church. And not a single stone has been laid to build a church in all of the Veneto country since he came here, and that is just the way that it is going to be.

"Likewise, should he ever rise to the papacy, not a single block of stone will be laid to the honor and glorification of God in the Catholic world until every single child in the world has a roof over his and her head and enough to eat.

"He justifies his refusal to build churches by explaining the difference between Christ and Moses' God to his congregation, *'The God of Moses is a self-serving God, one that requires His subjects to adore Him as the King He sees Himself to be - I am the Lord thy God and thou are to bow down to no other god than I. But, Christ, walks among His children, He wears no crown on His head, He carries no scepter, only compassion in His heart. He tells us, Love thy Neighbor as thyself - for I am thy Neighbor!'*

"The orphanage you passed at the foot of the mountain is a very special one, as it was the first one built in the world that was designed to allow children to be born who otherwise would be aborted by women too young or too poor to afford them. It is equipped with a world-class clinic and nursery which converts newborns into healthy two and three year olds, at which time they are placed into the community.

"Piccolo used the table money to break ground for the halfway house and clinic for the handicapped. He raised the balance of the money needed for that project via his popularity program; better known as his unpopularity program."

"His unpopularity program?" I repeated in a question.

"Yes, as I told you he is an enemy of the pomp and splendor of the clergy. When he came here, his first executive order was to ask the clergy to sell all their jewelry to build the clinic. Although most of the priests complied, a few of them sequestered their gold neck chains and rings of diamonds and rubies and emeralds. Nevertheless, you won't find a single solid gold cross hanging from any of their necks today.

"Not only the clergy got the shock of their lives when he showed up in Vittorio Veneto; the congregation did as well. Before he came here it was commonplace on Sunday to see as many as two hundred orphans congregating in the street that ran past the steps of the church; most of them were out-of-wedlock children and many others were emotionally or physically impaired, a few retarded, some missing limbs, others deaf or blind and of other deformities.

"Their only possessions were the tattered and torn rags they wore, mostly covered with mud in the spring and fall and frozen with flakes of snow in the wintertime. Then there were the tin cups they held in their hands, that is, those who were lucky enough to have hands. Most of the parishioners would sneak around to the rear entrance of the church in order to avoid them, similar to how Christians cross the street today to avoid the homeless in the street on their way to and from church.

"When Piccolo was named bishop and the time came for him to say his first mass, an overflowing crowd came from miles around to attend. When they showed up at the church on January 11, 1959, much to their surprise there wasn't an orphan to be seen. Some remarked *'It is about time someone cleaned up this mess.'* For the first time in years, they were able to ascend the steps and enter through the main entrance of the church. However, when they entered the church they got the biggest surprise of all.

"There on the end of each of the first few pews hung a small sign, *'Riservato ai bambini speciali di Dio'* - *'Reserved for God's special children.'* Lined up row after row were scores of orphans gazing at the magnificent Titian altarpiece. As I already told you, his first words to his newly acquired congregation were, *'Christ picked me up from the Mud in the Street and gave me to you.'*

"The *'Mud in the Street'* was a common expression used by priests and church goers in referring to the orphans of the time.

"Shortly afterwards, rumors spread throughout the Veneto country that the newly appointed bishop had been born out of wedlock. The rumor was fired by Piccolo's claim he had come from the *'Mud in the Street.'*

"Many priests called for his resignation from the priesthood and others went straight to Rome demanding he be defrocked. They had good reason for this, as it is *Canon Law* a born-out-of-wedlock child cannot be ordained a priest.

"Finally, when the situation started to get out of hand, the Archbishop of Venice published a copy of a marriage certificate in newspapers throughout Italy. The certificate evidenced his parents,

Giovanni Paulo Luciani and Bartolomea Tancon, had wed a year before his birth

"Nevertheless, since that time this good man has picked tens of thousands up out of the *'mud in the street'* and has given them their rightful place in society.

"Piccolo has performed a miracle here in the Veneto country. When he got here, the congregation thought all they had to do was go to church and drop a few coins in the box to be saved. Now they think much differently. Piccolo has created an army of people who are trying to help each other.

Gift of the Magi

"All over the western world, when a child reaches his or her sixth birthday, the child enjoys a big party. Everyone brings the youngster toys and other gifts. This teaches the child from the very beginning, gimme - gimme - gimme. How wonderful it is to get. We condition the child from the very start with greed; the *God of Moses* instruction in the Old Testament.

"Not here in Piccolo's world. On his or her sixth birthday, a child doesn't receive a thing. Instead, in the days leading up to one's sixth birthday, the child is busy making small craft items. On his or her sixth birthday, the child learns how wonderful it is to give. Everyone except the child gets a present. Christ's instruction in the New Testament.

"The child learns from the very beginning how much more wonderful it is to give than it is to get. Piccolo has tricked everyone here in the Veneto country into thinking Christ meant exactly what He had to say. In the past, each time the fork in the road would come up for each of them, they would ask themselves, *'Now what is in this for me?'* Now they ask themselves, *'Now what is in this for others?'* "

I started to take my seat but Jack waved me aside, "I thought I would give you the opportunity to sit in the same chair in which several popes have resided in the past. He picked the chair up and turned it upside down. On the bottom was a row of aging brass plates. On them were listed,

Leo XIII 11 November 1879

Pius X 2 January 1904
Benedict XV 24 June 1916
Pius XI 22 September 1923
Pius XII 2 February 1941
John XXIII 22 January 1959
Paul VI 22 July 1968

"Note the date for John XXIII is three days before he announced plans for *Vatican II* and change in the Church. In the following months he visited here quite often, spending as much as a week with Piccolo. It was difficult to tell which of them was the prodigy and which was the protégé. As history tells us, in that short time, John, one of the most conservative men in the history of the Church was converted into one of the most liberal men in the history of the Church.

"I questioned Piccolo as to how he was able to accomplish this transformation in the space of only a few months. He told me it took him only a few minutes. He told John what his father had told him many years before, *"Each time the fork in the road comes up, ask yourself, 'Now what would Jesus do in this case?'* The rest was history. After all, Christ was a liberal, you know.

"Other famous people have sat in your chair including Albert Schweitzer. He spent a week here five years ago. We didn't see either one of them from the time he arrived until the time he left.

"I asked Piccolo what they were doing all that time? 'We have determined that intelligence has nothing to do with genes. We evaluated, at great length, the testimony of Moses' God in the Old Testament and determined, from what we know today, He was a borderline idiot. On the other hand, we evaluated the testimony of Jesus in the gospels and determined that neither one of us could shine His shoes. The only other conclusion we could reach was that one was lying, and the other was telling the truth.' "

NOTE: For purposes of consolidation, the author relates his time with Luciani as if it was a single encounter, whereas there were more than a dozen sessions which took place on his visits to Italy in 1968 and 1969.

Chapter 22

The Winning Card

The footsteps coming down the stairs were obviously those of the man of the house. Momentarily he appeared, dressed again in street clothes unbecoming a bishop of the Church - a bright yellow shirt with ladybugs all over it, conventional slacks, and a pair of white bucks. He glanced at the table and looked at Jack as if he had said mass and had forgotten to serve the Eucharist. He slipped into the kitchen and returned with a frosted jug of water, which he poured into the water glasses that were on the table.

I have always taken the first sip of wine with bread so when the bishop started to pour the water I started to reach for the bread; suddenly from my left came a quiet kick. Sitting down and bowing his head, the bishop uttered a few inaudible words and I quickly bowed my head. Despite the fact he muttered in English, I had no idea what he was saying and wondered if he was reciting my grandmother's favorite version of grace, *"Father, Son and the Holy Ghost, he who eats the fastest gets the most!"*

The bishop offered a toast and Jack started, "Lucien is one of those who wants to be certain he's playing with a winning ticket."

Piccolo gave me a look of certainty, "I have a winning ticket for you, every time a winner, the winning card. If you promise to use it, I will give it to you. Guaranteed, guaranteed to win. No doubt about it."

Reaching for the truth, I asked, "Who is it? Is it Christ? Is it Allah? Is it Buddha? Is it Brahma?"

Much to my surprise he said, "None of those." Pointing to his temple, "It's here. All you have to do is do what it tells you to do. Always do what is right. Always do what this," again pointing to his temple, "tells you to do. That is enough. Nothing else counts. You do that, and that alone, and you're going to be in the winner's circle. Let there be no doubt about it."

"But, how can you tell which religion is right? There are so many of them."

"If you are strong enough to do what I just told you to do without any outside help, then that should be enough for you. For in that case you would be dealing directly with your maker. But if you need some help, some more cards, then you need this thing we call religion, the business Jack and I are in. And the winning card is not which religion is the true religion; it is what I have just given you, nothing more, nothing less. The purpose of religion, in fact the sole purpose of religion, is to help you play the game, and possibly help you play," again pointing to his temple, "the winning card."

He stopped and paused a bit and added, "I too had a mother and a father and today I know they are both in heaven. My mother was a believer, a devout Christian who prayed before crucifixes. The first thing I remember seeing was a crucifix she had tacked up over my crib. She saw in it the image of God and told me it was good and holy.

"My father, on the other hand, was a realist, one who sees things as they really are. He saw in the crucifix a near-naked man nailed to a tree, something a parent should not put in front of a child and tell him or her it is good and holy. He tore the crucifix off the wall and burned it in the stove.

"This does not mean my father did not believe in God. It was just that he did not believe in the business of religion. He could not accept that God would choose to appear to certain men and not to all of us. He knew that every single one of the so-called prophets had great motive to convince others they had talked to God. He was particularly suspicious of Moses and Joshua, who profited most of all by their claims they had talked to God. This made them the leaders of the Israelite nation and gained for them *'...all the gold and silver and brass of the thirty-three cities of the Promised Land.'*

"Nevertheless, since you raised the question: How does one prove Catholicism, Protestantism, Islamism, Buddhism, Hinduism or Taoism as the true religion? The answer is simple, one doesn't. It can't be done. It just can't be proved. Yet, what one can do is prove that a certain religion is not the true one.

"Let me give you an example of how this works; one cannot prove the right religion, yet, one can prove a particular religion is not the true religion," he took a sip of wine.

The expert said so; it must be true

"Sometime ago in Paris, where Modigliani had lived and worked, a sculptured marble head was recovered by divers from the canal that ran behind the artist's house. On the assumption the work may have been executed and discarded by the sculptor, it was brought to the attention of a local museum curator who believed it could be an original Modigliani. It was subsequently examined by experts in Paris and London who were recognized as the world's foremost authorities on Modigliani and it was found that, in fact, the work was an original masterpiece by the great artist.

"A local hippie challenged the find. The experts saw it as an opportunity to gain worldwide publicity for their discovery and arranged a live debate between themselves and the hippie on European television. The upcoming debate was widely publicized and the press made the hippie out to be some kind of a fool.

"The hour long debate allowed for a ten minute summary at its end for each of the parties. Although the hippie had disclosed several serious flaws in the experts' analysis, their immense credibility in the field of art enabled them to seal their claim as being sound. Art experts, like the preacher, have the advantage of unquestionable credibility, which totally eclipses anyone's ability to reason; anyone's ability to think. It was clear to all viewers the hippie really didn't know what he was talking about.

"Finally, the hippie got his ten minutes. He started by saying, 'I was hoping I wouldn't have to go this far but it seems I have no alternative.' Then he added something the art world has never since forgotten, *'No one,'* he told them, *'not all the experts on this stage or all the experts in Europe or, for that matter, all the experts in the world can prove any work of art is an original executed by a particular artist. On the other hand one can prove, beyond a shadow of any doubt, that a particular work is not authentic.'* To the astonishment of all, he proceeded to run a film of he, himself, designing and sculpting the alleged Modigliani head. He then closed with a single statement, *'The only way one can tell one is truly in possession of an original work of art is to watch the artist create it, watch him complete it, have him sign it and to take the work and place it in a safe to which only one knows the combination.'*

"Religions are like great works of art. When you believe one is the true one, as in the case of a great work of art, you are relying entirely on the credibility of the expert. You are relying entirely on the credibility of the preacher; someone who really doesn't know anymore about it than you do.

"The only way you can know a particular religion is the true one is to have been there. In the case of the Jew, this means to have been there on Mount Sinai when God the Father spoke His message to Moses. In the case of the Christian, it means to have been there when Christ rose from the dead and ascended into heaven. In the case of the Muslim, it means to have been there upon the winged horse when the Angel Gabriel took Mohammed to visit Abraham and Moses. In the case of the Mormon, it means to have been there at the edge of the pond when God the Father and Christ His Son appeared together to Joseph Smith.

"Otherwise you are relying entirely on what someone said to someone else, who in turn passed it on to someone else, who in turn passed it on to someone else, who in turn passed it on to someone else, for all time. You have to keep in mind, as my father claimed, in those days, as in these days, men had great motive to lie in these things; for if they were to convince their fellowman they had talked directly with God, it would make one a great and wealthy man.

"This includes Moses and Paul and Mark and Matthew and Luke and John and Constantine and Mohammed and Joseph Smith and all the others who claimed to have had talked with God. It also includes present day preachers who earn their livelihood this way, a few of which are trying to make this a better world to bring God's future children into; but most of which get their kicks out of wielding power over the minds of men. So much is the mind of man weakened by his mortality that he blindly pays the preacher for his salvation, despite the fact the preacher doesn't know anymore about the possibility of an afterlife than he does.

                    Other hippies run their films

"As with great art, it is often possible to tell a particular religion is not the true one. The most formidable enemy of the prophet is scientific fact and in some cases historical fact. We are seeing this happen each day more and more as we go on. As more and more scientific facts become available, more and more do we question the prophet and more and more is the credibility of the prophet eroded.

"The credibility of the art expert is similar to the credibility of the religious expert, the preacher. The lack of credibility of the hippie is similar, for example, to Darwin and Einstein in their day. In the case of the Modigliani scandal there are few people in the art world today who doubt the hippie's testimony. In fact, there is no one who doubts the hippie's testimony, because today one has the facts. But when he first made his claim, no one believed him, not a soul. For the credibility concerning the issue at hand was entirely with the so-called expert.

"Likewise, when Darwin and Einstein first made their claims, no one believed them, for the credibility was entirely with the preacher. The populace thought both of them to be insane.

"Darwin had proposed we had evolved from apes before the time of the first excavations of prehistoric bones of the early cavemen and the development of modern genetics which would later prove his hypothesis. Yet, Darwin's theory was based on things that could be readily seen; the general resemblance of man to the other higher primates.

"Even more ridiculous, at the time, was Einstein who spoke of things one could not see. He claimed all matter is made of moving parts called energy; including the hardest substance known, diamond. He made his claim before the time that modern microscopes and other advancements of science would prove his point.

"The marble floor we are standing on is made up of bits of empty space - atoms - which are traveling at such rapid speed they give the illusion of solidity. Our bodies are made up of atoms which are traveling at slower speeds, giving us the illusion of more porous objects.

"For Darwin and Einstein, the proof of the pudding was not to come about until after they had gone to the great house above. But when they first proposed their theories as to how we all came about, no one, not a soul, believed them. What they claimed contradicted what the experts were saying; what the preachers were saying; everyone knew it was a scientific fact God had created Adam and Eve to be the first parents.

"Like the hippie in the case of the supposed Modigliani, both these scientists have now had the opportunity to have run their films. Perhaps, after their time, but nevertheless, they have now been run. In the case of Darwin, archeology and more importantly genetics have proved his theory of evolution beyond a shadow of a doubt.

"In the case of Einstein, the development of immensely powerful microscopes and other scientific advancements have proved his theory of relativity. As compared with microscopes of Einstein's time, which had the capability to magnify up to seventy-five times, microscopes of today are able to magnify up to a million times. For those having no access to powerful microscopes, the splitting of an atom at Hiroshima should have done the trick.

"So now, like the hippie and his Modigliani they, Darwin and Einstein, too, have had their time on the floor. They too, have had their time before the cameras. What Darwin and Einstein once proposed as theory is now very much fact.

"The expert, the preacher, now has only two alternatives. He can admit he is wrong or he can avoid further questions. The only viable alternative he has, obviously, is the latter; to focus the attention of his followers on something else. He goes so far as to make it immoral, in some cases illegal, for his children to read these books. He takes these works of Darwin and Einstein out of his schools, out of his libraries; he does not want his children to know the facts. More so Darwin than Einstein, as the typical preacher does not have the intelligence to grasp the implications of Einstein's work.

"Einstein's *Theory of Relativity,* in its implications as to the origin of mankind, is much more devastating to the foundation of Christianity, the tale of *Adam and Eve*, than is Darwin's theory of evolution. After all, it proved the egg came first; God, as a matter-of-fact, did not create man and woman as adults. Einstein proved God's first creation was an infinitesimal speck which we now know as energy which gave birth to everything around us.

"The case of evolution is not quite as devastating, as the preacher has the loophole of claiming God created Adam and Eve as caveman and cavewoman. Some would go so far as to claim

God created them as the ape-man who later split into the higher primate ape and the higher primate man; the conclusion of modern science today.

"Nevertheless, the preacher causes his children to grow up shielded from what these great men had to say. Shielded from what are now known to be the facts, the hippie's film. He brings his children up in a world of make-believe.

"When Darwin and Einstein came along, the preacher would show up in the great arena, the courtroom, to debate the issue. Talking through his hat, armed only with his immense credibility, he would win every time.

"But where is he today? He is not there. He is not there because he knows the hippie now has his film, he now has the facts. The preacher now knows the hippie will make mincemeat of him.

"Religion, as I have said, is like a work of fine art. One might be an expert in religion. Another might be an expert in fine art. The preacher is the expert in religion. It is his job to win the credulity of his parishioners so that they will believe his teachings are authentic so they will respond with faith and ultimately with dollars.

"The art dealer, on the other hand, is the expert in fine art, and correspondingly, it is his job to win the credulity of his customers so they will respond with awe and ultimately with dollars. It's the same game, just being played with two different decks of cards.

"If the preacher is wrong and Christ, for example, is not the true emissary, then if the parishioner plays the card, I just told you of," pointing to his temple, "the card Matthew and Mark and Luke and John wrote of, he's still in the winner's circle. It makes no difference. On the other hand, if the artwork is not authentic, and the buyer enjoys it all his life, he's still a winner too.

"To answer your question more directly: No you can't prove which one is the true religion. I and countless others before me have tried. Faith must remain a product of belief, a guess at best. The important thing is that one uses the card I talk of, the winning card, and that he or she plays it throughout the game, every single day of his or her life.

"It may be the true religion is no religion at all. Keep in mind religion is a business that sells miracles and an afterlife in exchange for dollars. In short, it is mysticism. It may be the true

God is the God who we know as a matter-of-fact gives us life - the *God of Nature* - the God of the atheist. Common sense tells us children who grow up outside of organized religion make better members of society as they grow up with fewer prejudices.

"It is that my father was an atheist that made him a true follower of Christ. My father knew if Christ actually performed the miracles said of Him, He would be a historical figure like many other common men like Edison and others have left their miracles behind them. But, none of Christ's miracles have survived to this day. It was that my father saw Christ as a man, and not as a God, that enabled him to see past the mythology of Christ to the reality of Christ.

"Although he did not believe a man ever lived who had performed the miracles that are said of Him, my father did believe good men had written the gospels; men who wanted to do away with the evilness Moses had left behind."

The bishop paused a long time as to give each of us time to catch up. He nailed the lid shut.

"On Napoleon's tomb is engraved the words, *'Man's only immortality is what he leaves behind in the minds of men.'* This is true not only of men, but also of Gods."

Both Jack and I gave him a look of apprehension. "It makes no difference whether or not Christ ever lived or whether or not He performed the miracles that are said of Him. What is important of any man, even one who claimed to be God, is not so much His life on earth but rather what He left behind; like Tao, like Buddha, like Mohammed, like Lincoln, like Einstein, like Edison and all the others who have come before and after Him.

"What is important of Christ's life is not His life on earth, not the miracles He is said to have performed, not His death on the cross, not even His ascension into heaven. Like any God, or for that matter, like any man, all that counts is what He left behind,

'Love thy Neighbor as thyself . . . all creatures, great and small.'

"We must work together toward a time in which all of God's children, no matter how scorned by man-made doctrine, will be accepted with equal human dignity under the laws of nations."

He carefully studied our expressions to assure himself we would remember this last comment for all the days of our lives.

He asked me. "Where do you go from here?"

"To Rome, St. Peter's. Then, back to the states."

He offered, "I could make arrangements for you to stay in one of the Vatican apartments, but I don't think you would feel very comfortable there. Not far from St. Peter's is a little pensione. You might feel less ecclesiastical there." He took out a card and wrote an address on the back of it and handed it to me.

He retrieved a second card and wrote something on its back. As he handed it to me, he said, "When you visit the Vatican find the office of Bishop Marcinkus. It is in the Government Palace which is the largest building in the Vatican gardens. He was here last week. We were fighting over the branch of the Catholic Bank. He intends to turn it over to private interests, something I don't agree with, but nevertheless something that is going to happen.

"He owes me one. I am sure he will be happy to arrange to have someone give you a personal tour of the gardens; maybe a walkthrough of the Papal Apartment itself. You will find this quite interesting as it something the public never sees.

"Thank you," this man with the perpetual smile added, "You have enlightened me. You have given me good substance for my sermon on the mount." Getting up he stopped at the sideboard and picked up a few pieces of stationary headed *Vittorio Veneto.*

"I will make notes of what we have talked of today for posterity. Shot for the cannon. I will see you on the battlefield." He headed for the stairs. I followed him with a puzzled look.

"The enemy," he shot back, "politicians and preachers who prey on the weakness of the minds of men."

Smiling at him, "You can count on me, I will be there."

As the bishop's footsteps faded up the stairs, Jack offered, "Forgive him. He retires at precisely nine o'clock and is up at four. Even if the Pope was here, he would do the same thing."

"Wow, up at four. He must sleep all afternoon."

"Wrong. He has no idea what a nap is. Come, I will walk you to your hotel."

It was sheer silence as we walked through the town. I broke the silence. "What do you think his chances are at the papacy?"

"Although he is careful not to advertise it, he tells others he has no interest in it, it is his driving ambition in this life to accede to the papacy. Because of his work here at the school, he is a recognized authority in theology. Something he calls 'human theology.' As a matter-of-fact he has been a chief advisor of both John and Paul during their papacies. He's well liked, probably a few years from being named a cardinal, an Italian Cardinal. He has a good shot at it. Yet, he wants this not for the vestments and the pageantry. In fact, he tells me if the day ever comes he will refuse the Papal Tiara which has been the focus of coronations ever since popes have been elected.

"He wants to be Pope because he will be able to change things for the better. He wants to complete Patton's dream. He wants to bring about the equality of all God's children. Patton, with his army, would have given us the diamond and Piccolo, with another kind of army, will make it real.

"Patton, with his army, has won for us *freedom* and Piccolo, with another kind of army, will win for us *equality;* something Patton could not do. *'For Freedom without Equality is not what it pretends to be, the diamond would be made of paste.'*

"Last year, Paul VI visited here. The next week he issued the decree banning birth control practices in the Catholic Church. The very next week, Piccolo issued an editorial to the press calling for repeal of the doctrine.

"In retrospect it seemed to me this strange order of events had been planned by the two of them. Certainly, Paul's ruling was not consistent with his own thinking on the subject and ignored the conclusion of the committee he had appointed two years ago to study the issue.

"It seems to me, the two of them decided Paul would issue the proclamation and shortly afterwards Piccolo would follow publicly with his objection. Perhaps, history will someday tell us why they did these confusing things. One thing I do know is that shortly afterwards, Piccolo's attitude changed quite dramatically, from one of wanting to be the next Pope, to one of knowing he would be the next Pope."

As we reached the hotel he hugged me goodbye and that was the last time I ever saw him.

Chapter 23

What They Died For

As the plane passed over the harbor, I looked down and yes, she was still there. Now, I knew why she was there. For she knew what she was doing when she cried with silent lips, *"Give me your tired, your poor, Your huddled masses yearning to breathe free, The wretched refuse of your teeming shore. Send these, the tempest-tost to me. I lift my lamp beside the golden door!"*

She didn't care if they were Christian or Jew, or Muslim, or Hindu, or Shinto, or Buddhist or Tao. For she knew she must gather from the ends of the earth the great dispersion of minds needed to accomplish the great task which lay before her; to mold America into what it is today.

She didn't care whether the men she needed to build and light her cities were Spanish or Dutch. She didn't care whether the men she needed to build her steel and paper mills were Russian or Scandinavian. She didn't care whether the men she needed to build her roads and bridges were Polish or Irish. She didn't care whether the men she needed to establish her communication infrastructure of telephone, radio, television and computer technology were Italian or Chinese. She didn't care whether the men she needed to get her airborne were French or English. She didn't care whether the men she needed to bring her into the nuclear age were Swiss or German. As a matter-of-fact, she didn't even care if they were men or women.

Above all, she didn't care whether Private John Doe was Christian, Islamic, Jewish, Hindu, Shinto, Buddhist, Tao, atheist, black, white, red, yellow, straight or gay, when, in his foxhole he wrote,

*"I don't know why this terrible war is being fought,
I don't know what will be its outcome.*

*"But, what I do know, is that I plan to fight this war as if
its entire success or failure depends on me alone!"*

Distinguished Service Cross for valor not quite justifying the **Medal of Honor**

Chapter 24

Thoughts

Later that year, Piccolo became Archbishop of Venice and Jack was transferred to Milan where he served under Cardinal Pignedoli until 1975 when he was moved to the Vatican and assigned to the *Council for Spiritual Occurrences*.

In 1978, when his patron Albino Luciani was raised to the papacy, Jack was assigned as an intermediary between the Pope and the Vatican cardinals with the job of quelling the uncertainties of those who were expected to lose their positions.

Jack's death was coincidental to that of his patron. On the second day following the Pope's unexpected death, Jack was killed by a hit-run driver just outside the Vatican walls. The driver of the car has never been apprehended.

... halfway between a Silver Star and a Purple Heart

As I drove into the funeral home parking lot, I thought of my last visit with Jack, one of the few times I had seen him in the twenty years since the day he had stood at the pinnacle of secondary school life rendering the coveted address on graduation day. Perhaps he should have used the words Lincoln had used so effectively, so modestly, at Gettysburg, *"... the world will little note, nor long remember what is said here..."* for unlike Lincoln, Jack would have been telling the truth.

I thought of Jack's letter in which he had told me of John Paul's private meeting with the *Curia* cardinals just a week earlier, *'Mother Church will cease to be the cause of many of the world's problems and rather will begin to be the answer to them.'*

I thought of this great man Piccolo who lay in state on the grand catafalque at that very moment, partway around the world. Greatness, I thought, does not always give notice. Like that of his patron, Jack's was a simple, quiet, silent greatness; one whose words would not long be remembered, and whose deeds would long be forgotten. But nevertheless, one whose purpose would live on for all eternity, live on in his fellowman.

As I stood before him, I couldn't see him, for the mutilation had been much too terrible to allow the body to be viewed. As I began to realize he was now still, I recalled what he once was and I prayed he had had time to have completed his work. At the same time, I thought of Piccolo and I knew he had not had time to complete his work; which has caused me to write this book.

I reached over and lay my hand on the lid of the coffin, just above his heart, to confirm to myself that this was indeed forever goodbye. One of those things one calls tears came up out of my heart and started from the crevice of my eye and crept toward the lid. I glanced first to the right, and then to the left, and again to the right, and finally to the left, once more. I carefully held it there balanced on the edge of the cliff.

Turning to the audience which sat in grief, I heard applause move forward in muffled cries; in sighs of desperation and sobs. I went directly to the one who sat in the front row and introduced myself and told her of a part of my life which had been a part of her son's life.

She reached into her purse and brought forth a package and placing it into my hand she said, "Jack told me if anything was to happen to him he wanted you to have this. We gave it to him for his ordination, the most important day of his life."

Later in the funeral home parking lot in the privacy of my car, I looked at the package. It was still wrapped in its brown paper covering. It was postmarked *Centrale Poste de Roma* and dated, September 23, 1978, a few days before John Paul's death. As the package had never been opened, Jack had undoubtedly called his mother sometime between then and now.

I was puzzled why it was not postmarked *Poste Citta del Vaticano,* as had been true of Jack's correspondence to me through the years. I unwrapped the package and opened the small box within it. I saw it was six o'clock, precisely halfway between a *Silver Star* and a *Purple Heart.*

And Piccolo? Well that was the last I saw of him, too. Yet, I have never forgotten him. I will always remember him. He will always be in my thoughts. Partially, because of his unrelenting almost *Mickey Mouse* smile, but mostly because of his toes.

Epilogue

Imagine

The following is a *What if?* satire; what might have been had his life not been cut short? *Italicized* text, however, is taken from John Paul's ministry and other real people and is not original to this text. These quotes are footnoted elsewhere in this book.

Associated Press, October 12, 1978 Vatican City. Pope John Paul has summoned the one hundred thirty-eight cardinals and a countless number of bishops and other ranking prelates of the Church to Rome. Although the Vatican was mum concerning the Pope's action, rumors quickly surfaced, John Paul would resign his pontificate, the first pope in the two-thousand year history of the Church to do so. It is believed the Pope is unhappy in Rome where he is surrounded by the *Vatican Curia* which, it is reported, is strangling his papacy; something his predecessor put up with for fifteen years.

London Times, October 16, 1978 Vatican City. Church leaders from the ends of the earth poured into Vatican City today. With them were hundreds of reporters representing every major newspaper and television station in the world. John Paul remained sequestered behind closed doors in his private rooms preparing his resignation speech. His address will be televised worldwide tomorrow from St. Peter's Basilica.

London Times, October 17, 1978 Vatican City. Reliable sources in the Vatican report Pope John Paul submitted his resignation this morning to Secretary of State Jean Villot. Villot, as the interim Pope, will most likely lead the formal resignation ceremony tomorrow. Frontrunners for the succession are Cardinals Benelli of Florence and Siri of Genoa . . .

Associated Press, October 18, 1978 Vatican City. Rumors of a papal resignation went up in smoke yesterday at a special assembly of the *College of Cardinals* and other ranking prelates of the Church. Garbed in a plain white smock and standing without the aid of a podium in St. Peter's Basilica, John Paul rocked the Roman Catholic world in announcing plans for *Vatican III*.

More than five hundred cardinals and bishops dressed in similar white smocks, as one would buy in a Job-Lot, sat stunned in their seats as the leader of the world's largest congregation opened his address with what had been the central message of his acceptance speech a few weeks earlier. *"We will bring about a new order, this one more just and equitable for all God's children. We will rise up the courage that is within us and set aside the convictions of our Christian forefathers; together we will muster the strength to lift those restraints that have been unfairly placed upon the everyday lives of so many innocent people by doctrine... for God-given human life is infinitely more precious than is man-made doctrine..."*

John Paul spoke of the problems of the world. "We must accept it is the Church's irresponsible ban on contraception that is the driving force behind the spread of disease, poverty and starvation in third world countries and abortions in first world countries. What's more," he told them in a dictatorial tone, "we will do something about this; and we will do it now."

"In the real world in which we live, the rich nations of North America and Europe are ignoring Rome's edict; the people themselves are taking upon their shoulders to control the world's population. The Catholic population, discounting immigration, in these countries has been declining. Yet, the population in Latin America, Africa, and other impoverished parts of the world has been skyrocketing; these Catholics have been blindly following the Church's directive to build large families. So, yes, the overall Catholic population has been growing, but this growth has been generating poverty and starvation in underdeveloped nations."

The Pope continued, "My good friends, do you not know what it is to starve to death; particularly, as a little child? The pain and suffering is immensely more excruciating than that of Jesus on the cross. Does this not bother you each day when you pray before marble statues in the golden grandeur of your churches, which at auction could buy healthy lives for tens of thousands of these children? When the robes and gold you wear could be traded in for a thousand more? Do you not remember the hallowed words of Cardinal Gantin of Africa when he wrote to Paul, *'I pray for a day when all children are brought into a world of good health, happiness and opportunity, yet, one must face reality that today we bring many children into a living hell.'"*

Speaking to a Catholic world television audience, the Pope repeated much of what he had to say in both English and Spanish, yet, on occasion, he would repeat certain phrases in more than a dozen languages. *"What is important is not how many children are born. What is important is that every single child that is born has an equal chance at a good and healthy life.* Whereas many of the things we will speak of today can be distributed to committees or councils for discussion and examination and proposal for a period of months or years, this one cannot wait.

"If the information I have been given, the various statistics, if that information is accurate, then during these few minutes I have been speaking, over one thousand children under the age of five have died of malnutrition. During the next few hours, while you and I look forward to our next meal, five thousand others under the age of five will die of malnutrition. By this time tomorrow, thirty thousand others under the age of five will be dead of malnutrition."

The Pope continued to speak in strong authoritative tones, *"There are more than one hundred mullion children alive today who will never see their tenth birthday. And the reason they will never see their tenth birthday, is the Roman Catholic Church. God does not always provide. It is our sacred duty to provide. And we will provide now.*

"Mother Church will accept her responsibility to better manage the world population. She will cease to be the cause of many of the world's problems, and instead will begin to be the answer to them.

"In this thing I need no advice. I need no authority other than that given me by Christ," the Pope pointed to his temple. "At midnight tonight, according to the power vested in my office by *Canon Law*, the doctrine *Humanae Vitae* will cease to exist. Catholics all over the world will be free to practice *Planned Parenthood* as their individual circumstances and consciences permit. Those outside of marriage will be encouraged to use appropriate contraceptive means to prevent conception. We must all join together to bring about a day when every single child that is born has an equal chance at a good and healthy life."

A shudder ran through the assembly. Fearful expressions on a field of faces foretold what would come next. The Pope continued to take advantage of his dictatorial powers in just about everything he had to say. "Unfortunately, lifting the ban on contraception will not help the hundreds of millions of children who are, as we speak, starving to death. For this, we must make sacrifice. And, again, we will make that sacrifice now.

"In my first executive action two months ago, I ordered a complete audit and valuation made of the Church's worldwide assets including the Vatican treasures. Again, if the numbers the appraisers have given me are correct, and I have every reason to believe they are correct, the Vatican treasures would bring at auction not millions, not hundreds of millions, not billions, but tens of billions of dollars. The Pieta alone and several other individual works by pinnacle artists would bring in excess of a billion dollars each. I have the record here before me that the Vatican treasures stored in warehouses - not even on display - would bring billions by themselves.

Michelangelo's *Pieta*

"There are some who will say liquidation of the treasures of the Sistine Chapel and the Apostolic Museum will lose for the Church the vast revenues obtained from tourists. Yet, the truth is, the gross tourist revenue for ten years does not pay for the maintenance of these old buildings for a single year.

"Nevertheless, we are talking here of God's treasures; those we hold in trust here in the Vatican for the Almighty *God of Moses* who promises us the Kingdom of Heaven for not much more than we fall down on our knees and adore Him."

His listeners squirmed in their seats as John Paul reminded them of Christ's instructions. *"Not once in His ministry did Christ tell us to fall down on our knees and adore Him. Not once did He hint at this deception we as His representatives here in Rome play out on the world stage. We have become ourselves, the hypocrites – the Pharisees - Christ so often condemned.*

"Believe me, when I tell you of Christ's message, *'The Kingdom of Heaven lies not in magnificent buildings of wood and stone; nor in chalices of gold or priceless art; nor in chants and praises... it lies in our hearts... our compassion for others.'* Christ tells us many times, building magnificent palaces of worship has nothing to do with salvation.

"The Vatican treasures are not limited to durables. The Castel Gandolfo also boasts one of the world's finest wine cellars. I am told that when a vineyard sends a bottle of wine to the Vatican, it sends only the very best; thus even though we don't want to, we must drink the very best." He nodded at a nearby table on which sat a bottle of wine. "I am told this bottle alone would fetch several thousand dollars at auction."

papal lunch

Pausing between each of a dozen languages to emphasize his next point, the Pope spoke in a way as if he was addressing each member of his worldwide audience individually, *"Christ gives each one of His children at birth the key to the Kingdom of Heaven - one's compassion for others. Let us not be so foolish as to cast it aside for Satan's offering of gold."*

John Paul quoted his own remark of two weeks earlier at a general audience, *"'... One day, we who live in opulence, while so many are dying because they have nothing, will have to answer to Jesus as to why we have not carried out His instruction, 'Love thy neighbor as thyself.' We, the clergy of the Church together with our congregations, who substitute gold and pomp and ceremony in place of Christ's instruction, who judge our masquerade of singing and dancing His praises to be more precious than human life, will have the most to explain.'"*

The Pope again repeated in several languages, "There is a story of my childhood I have told and retold many times. So often, I cannot bear to hear it again, myself. Yet, I have never told it at a more meaningful point in my life than now.

"When I was teenager, my father made me promise I would live my life in Imitation of Christ, and I have kept that solemn promise. Each time, the fork in the road has come up for me, often only minutes apart, I have asked myself, 'Now what would Jesus do in this case?' I have often pondered the possibility how much better the world would be if everyone were to do this.

"Now, the fork in the road has come up for us. To the *right*, is the promise of the *God of Moses* of the Kingdom of Heaven for not much more than we store up treasures on this earth to His glory and fall down on our knees and worship Him. To the *left*, is Christ who promises us the Kingdom of Heaven on the sole requirement *'Love thy neighbor as thyself.'"*

The Pope continued to speak each word with absolute conviction, "We will choose between the *God of Moses*' treasures - gold, priceless art, feather pillows and two hundred dollar bottles of wine - and Christ's treasures - the tens of millions of children who are starving to death throughout the world as we speak. We will choose between the selfish word of the *God of Moses* and the compassionate word of Jesus Christ, Himself. We will make that choice. We will make it now.

Opus Dei's Church for the rich   Luciani's Church for the poor

John Paul repeated his statement of September 27, 1978,

"'. . . One day, we who live in opulence, while so many are dying because they have nothing, will have to answer to Jesus as to why we have not carried out His instruction, 'Love thy neighbor as thyself.' We, the clergy of the Church together with our congregations, who substitute gold and pomp and ceremony in place of Christ's instruction, who judge our masquerade of singing and dancing His praises to be more precious than human life, will have the most to explain.'"

Author's note: More than anything else, this is what Albino Luciani was about; to change the Church back to what Christ had intended. More than anything else, this is the effort of this book.

    ABOVE: One of 22 papal tiaras encrusted with 1,700+ diamonds, rubies and other precious gems. The Vatican owns more than 5,000 golden chalices; the one held by John Paul II boasts 122 diamonds ranging from 3 to 27 karats. The Vatican wine cellar contains more than $20 million in vintage wines. The Vatican owns the largest collection of ancient sculpture in the world and its works by pinnacle artists would fetch tens of billions at auction. Christ actually gave us two commandments: 1) *Love thy neighbor as thyself* and 2) *Sell all thou hast and give to the poor.* Today, the papacy remains the most powerful force on earth against a redistribution of wealth society and its ban on contraception drives disease and starvation. It embraces Christ's two great enemies: *GREED* and *BIGOTRY;* it refuses to accept the equality of woman, homosexuals and others. It fools its congregation into thinking adoration of the God of Moses is the key to salvation.

"Mother Church is about to carry out Christ's most prolific instruction, *'Sell all that thou hast and give to the poor.'* She is about to shed her hypocrisy and show the world her heart. There is no time here for discussion, no room for bargaining, not a moment for quibbling, not a moment of delay. We will act together. We will act quickly. We will act now."

Vatican III

Had it not been for the rumors of resignation, what the Pope had to say in the morning session - although horrifying to those on the conservative *right* - came as no surprise as he had been hammering away at the Church's irresponsible role in the world's population problem and the hypocrisy of the Vatican treasures for twenty years. What occurred when they reconvened in the early afternoon must have shot thoughts through their heads as to how they could possibly have him committed.

The Pope opened the afternoon session, "We must take the time to define our God." Murmurs rumbled through the audience. Many shot questionable glances as one looks at a madman.

Andrew Rublev icon

Angels representing the *Holy Trinity*
discuss the mystery of three persons in one God

The Pope set forth his plans to redefine the definition of the Christian God as it had been determined by the *Nicene Council* in 325AD. The Council produced the Nicene Creed - today's Apostles Creed - the founding document of Catholicism - the pledge of allegiance to the Catholic Church. The creed defined Christ as a manifestation of the *God of Moses*; they were one and the same God. The Pope continued.

"In the real world in which we live today, we can no longer pretend to play with men's minds. We can no longer pretend to claim Christ was simply a manifestation of a God who believed in the superiority of man over woman. A God, who told us some of His children were better than others, and others were to be subordinated, cast into slavery and in some cases annihilated.

"In addition, as intelligent human beings, we must accept what we know to be the truth today; the facts science and history have determined for us. We can no longer pretend to fool our followers with a facade of fairytales and mysticism, for eventually our credibility will fail and with it Mother Church will fall.

"The young men and women of tomorrow will say, 'If these men in Rome insist that Moses and the Israelites were in Egypt for four hundred years and Moses did, in fact, part the Red Sea and he did, in fact, take the Ten Commandments from his God on Mount Sinai, a tablet we know today was written by a man of less than average intelligence and organization, much less a God; why should we believe in anything else they have to say.'"

John Paul paused a moment and asserted his next point, *"Children, who today lose their Santa Claus at the age of six, will soon lose their Christ at the age of seven."*

He paused again, *"Now listen to me carefully."* He waited for each ear to perk up as to not miss his next point. *"I do not believe the same God who endowed us with reason and intellect intended we forego their use in determining the truth. The truth does not lie in the past. It lies in the future. It lies in what we have determined to be the truth to this day and what we will determine to be a better truth tomorrow."*

The Pope stopped again. This time speaking more slowly and even more firmly, he told the assembly, "Every child of six knows the world is round. Let us not be so foolish as to tell them we believe in a God who thought it was flat." Again, he stopped as to give time for comprehension. "Every child of six knows the inhabitants of the Americas first walked across the ice-bridge from Asia twelve thousand years ago. Let us not be so foolish as to tell them we believe in a God who created the first man and woman six thousand years ago."

He paused again as to give time for the slowest of his listeners to understand what he had to say. After all, they all knew God had created Adam and Eve six thousand years after the Eskimos first walked across the Iberian Straight. He wasn't telling them anything new. To most of them, he was confusing belief with reality; in their minds one had nothing to do with the other.

John Paul, surrounded by an ocean of incomprehension and doubt, paused long enough this time to give them some measure of hope he would change the subject. After all, the world's children were witness to what he was saying. He would do irreparable damage to their faith. Instead he gave them more of the same.

"The world outside no longer accepts what the *God of Moses* left behind. The world outside no longer believes in slavery, the superiority of the white Aryan male, the servitude of woman to man, the segregation and annihilation of subordinate races and the persecution of those who live their lives differently. Also, it is coming to know sex as a great and beautiful gift from God, rather than the vile offering of Satan, Moses would have one believe. The world outside is changing. It is coming to accept the truth. We, too, must change. We, too, must come to accept the truth.

"Conversely, the world is coming to accept what Christ left behind more and more each day, *'Love thy neighbor as thyself.'*

"Christ is a God for everyone; believer or nonbeliever."

John Paul clarified his point, "It makes no difference whether or not He ever lived. It makes no difference whether or not He performed the miracles He is said to have performed. It makes no difference whether or not He was born of a virgin, died on a cross, rose from the dead or ascended into heaven. Like any other God, for that matter, like any other man, all that is important is what He left behind, *'Love thy neighbor as thyself.'*" John Paul raised his voice an octave or so, "The time has come for all of us, each one of us, to shed our hypocrisy and get up off of our knees and do what He told us to do."

The Universal Christ

Again, the Pope broke for a moment or two to allow the most sluggish of minds to catch up with him. "That Christ is the true God is the witness of all religions of the world. Christ's most prolific testimony, *'Lay not treasure upon this earth where moth and rust doth corrupt but lay up treasure in Heaven'*... *'It is easier for a camel to pass through the eye-of-a-needle than for a rich man to enter the Kingdom of Heaven'*... the parable of the widow's mite in which she gives her last pence to the poor... *'If thou wilt be perfect go and sell all that thou hast and give to the poor and thou shalt have treasure in Heaven.'*

"Repeated over and over again, in as many different ways, as if we might not understand it in one way we would understand it in another, are Christ's sole requirements for the Kingdom of Heaven, *'Love thy neighbor as thyself'* and *'Sell all that thou hast and give to the poor.'*

"Christ's message is consistent with Mohammed's *Koran* in the Muslim world, *'Give all that thou hast to your neighbor. For I am your neighbor. If the wealth ye have gained, and merchandise ye fear may be unsold and dwellings wherein ye delight, be dearer to you than God, dearer to you than your neighbor, then God too will be dearer to others. And you will not reside with Him in His house this day, and for all time.'*

"On the other side of the world is the ancient scripture of the Hindu world - the *Vedas*. The word of the God Brahma, *'Care only for others, less I will not care for you. For I am the others. This is all I want of you. For you will not see the brightness that lies beyond the shadow of doom unless you care for others. Unless you care for me.'*

"In China, Tao, too, in all of his wisdom, five hundred years before Christ, shared this same fundamental philosophy, *'Love others as you love yourself.'* and *'He who stores up gold in his temple for his God in this life has had his reward in this life. He who sells his gold in this life to help others will have his reward in the next life.'*

"Buddha, before Tao, in his *Tripitaka* left this very same message behind, *'When all is said and done there are only two measurements of life. Oneself and others. The first will lead to treasures in this book of fools. In this book of this life. The other will lead to treasures in the true book. The vast book of eternity.'*

"Christ's message is also the heart of the *Bible of the Atheist* - the so-called nonbeliever - who uninfluenced by the hatred of preachers through the years has grown up free of prejudice toward his fellowman. Unlike we who navigate our way through life immersed in organized religion - organized hatred - he has only Christ to guide him," the Pope pointed to his temple.

"The Universal God - Christ - has etched His message in the Bibles of all peoples of the world. He has provided them with an equal chance at eternal life. The rich man who tosses his pennies into the poor box is passing up that chance. We who value man-made doctrine, golden treasures, pomp and ceremony to be more precious than human life are passing up that chance."

He speaks to imbeciles

The Pope continued to nail down his position, "If we keep insisting Christ is the same God as is the *God of Moses*, sooner or later our congregations will wake up and ask themselves. 'If Christ is this same God, then it was Christ who destroyed the world together with all its infants and children for sins of the flesh in the story of Noah's Ark. It was Christ who murdered the infants and children of the defenseless Canaanites in the horrific taking of the Promised Land. It was Christ who commanded children born as little girls, like animals, are man's property and are to live their lives in dire servitude to their male masters. It was Christ who commanded out-of-wedlock mothers and prostitutes be taken outside the walls of the city and stoned to death. It was Christ who gave us the Tenth Commandment which protected the right of one man to enslave another. It was Christ who commanded children born with flat noses (Negroes) and children born handicapped; specifically blind children, deaf children, crippled children, dwarfed children, children born with countless deformities and children born-out-of-wedlock are not worthy to approach the presence of the Lord.'

"They will ask themselves, 'If those men in Rome believe it was Christ who did these horrific things then why should we believe anything else they have to say? It is obvious to an imbecile, Christ could not be the same God as was this diabolical creation of the mind of Moses.

"History, archeology and science have eroded just about everything Moses had to say. Yet, on the other hand, these advancements have no contest with what Christ had to say.

"Those in the outside world, more and more each day, are setting aside what Moses had to say, and are coming to accept what Christ had to say. Let us not allow our cowardice - our fear of the *God of Moses* - to cloud our vision for the truth - Christ."

The *Holy Trinity*

The Pope began to take on the stature of a trial lawyer, "We are in the business of religion. We are the wardens of faith in society. We claim to be the 'experts' in this thing one calls 'God'. Yet, we speak of 'God' as a *Holy Trinity*. Something we say we don't understand. We gamble our eternity on something we don't understand. We assume God, who has endowed us with reason and intellect, intended us to forego their use in defining our God."

He defined the mission of *Vatican III*. "It is time for us to learn our business. We speak of God everyday. We are talking about God right now. Let us put on the table precisely what we are talking about. Is our God a God of hatred and fear? Is our God a God of love and compassion? We will make a choice. And we will make that choice now." he told a sea of confused white smocks.

The greatest of sins

Having defined the scope of *Vatican III,* John Paul immediately moved on to other issues. "When I was a little boy," he told them, "sex was an unmentionable. After all, Moses had taught us sex is the worst of sins." It came as no surprise when he repeated the most prolific testimony of his ministry, *"We have made of sex the greatest of sins whereas in itself it is nothing more than human nature and not a sin at all."*

As he had done in the case of his other declarations, he repeated it in a dozen different languages so that all could bear witness to his testimony. Though he had said it hundreds of times over and over again as a young priest, as a bishop and as a cardinal it had been nothing more than opinion. Now, he had said it as Pope. It was *Canon Law*.

"Today, we are coming to know sex in itself is good and beautiful, a great gift from God. It is a matter of very personal intercommunication between two people who love each other and very often it rewards us with the greatest gift of God, a newborn child. It is a private matter and should not be the business of a group of old men in Rome who grew up thinking there is something wrong with it."

The Pope alluded to *Canon Law*'s condemnation of teenage masturbation, "It is particularly wrong of men in the Vatican, who have been unable to accept the innocence of sex themselves, to condemn a child for growing up - in the way God deems he or she grow up." Again, he raised his voice as to leave no wiggle-room for dissension, "Mother Church will accept Mother Nature. We, too, will accept Mother Nature. We will accept her now.

"We will begin today to cease concerning ourselves with what people do in the privacy of their bedrooms which should be none of our concern. We will instead turn ourselves to the real problems of the world which should be our concern."

The Pope talked briefly about the confessional box. "I want to clear up a rumor I will do away with confession for children and teens until they reach adulthood. Not true. Yes, the confessional box will no longer be a haven for discussing one's sex life. Yet, it will continue to be a place for those guilty of betrayal, murder, rape, theft, molestation and so forth. But, most importantly, it will become a haven for love.

"I am quite dismayed that in all my priestly life, I have rarely been approached by a parishioner asking forgiveness for the greatest sin of all, hatred. This includes the confessions of many priests who I know openly preach hatred of different kinds of people going so far, in some cases, as to take part in political agendas which deprive them of equal rights under the law. They don't think there is anything wrong with it despite the fact it breaks Christ's greatest commandment, *'Love thy neighbor as thyself.'*

"Believe me, there is no room for hatred in Heaven. A man, or even a child, who dies with hatred of any kind in the heart whether it be of people of other races, creeds, ethnic origins or even of homosexuals or atheists, goes straight to hell." It was obvious from the shocked looks on their faces the Pope had struck a nerve of shattering proportions in most of his listeners.

"We will begin to preach this greatest of sins. It will be the central focus of the confessional box. Children, of all ages, who develop the slightest tendency of hatred, will know they must go to confession. The priest will no longer give the child vain repetitions of penance, three *Our Fathers* and ten *Hail Marys,* but instead will take the time to give them corrective counseling.

"*Organized religion* is about to cease to be what it has been for centuries, *organized hatred,* and instead will begin to be what Christ intended it to be, *organized love."*

A world of perfect children

Taking care not to fill the afternoon with rhetoric, the Pope, having made his point, moved on to his next message; his final statement of the day. Speaking more slowly and precisely now, still his voice resonating throughout the immense church, even those who were sequestered in the crypt below could hear each word. Outside, his unyielding voice carried throughout St. Peter's Square and to the ends of the earth. The Pope spoke of the world's future children.

"Two months ago, the world's first artificially inseminated child was born in England. I sent a message to her parents, *'I congratulate you on the birth of your little girl. I have no right to condemn you for what you wanted and asked the doctors to carry out. Be assured, there is a high place reserved for both you and your child in Heaven.'* I know this caused much distress for many of you. I want to take the time to explain why I sent that message."

The Pope stopped again, took a sip of water, and went on, *"My good friend Einstein once told me that he could not accept the existence of God because he could not accept that God would play dice with His children, that there is something horrific about how God goes about making children, that millions of fertilized eggs are cast into the sewer in everyday intercourse, and many others are destined only to be born and die unspeakable deaths.*

"Here on the eve of the realization of this great event in genetic research, from a practical point of view, certainly from a

humanitarian point of view, it makes no sense for man to allow God to continue to play dice with the world's children.

"The question we have to ask is: Is it right? Is it right for the child to suffer because of the insane ideology of the preacher who thinks what someone is supposed to have said to someone else thousands of years ago determines what is right and wrong today. The answer is clear to me," John Paul paused and again repeated the phrase in a dozen different languages, *"What is important is not how many children are born, but that every child that is born has an equal chance at a good and healthy life.*

"I believe genetic research will eventually lead to man's greatest achievement, the creation of a perfectly healthy child every time." The Pope raised hairs on the backs of their necks, "The day is not far off when all children will be born of artificial insemination and natural conception will go by the wayside. I am speaking here not of centuries, but of the not too distant future.

"This does not mean the child will not be God's child. This I see as the bottom line. Not that it is my child, or your child, or his child, or her child; all that counts is it is God's child." He paused once more as if to insure himself they had not missed his point.

"We are now only decades away from the time doctors will be able to test the sperm and the egg before conception to weed out genetic impairments, disease and other birth defects to make possible a perfectly healthy child every time. Responsible parents-to-be will choose artificial insemination over natural conception to guarantee the health of their offspring.

"We will see this first applied to eliminate transmission of diseases in pregnancies. The present plague of Ebola, which is currently rampaging across Africa, comes to mind, but there will be other diseases in the future; perhaps, some even more deadly. Eventually, people will come to realize children born of artificial insemination are not only born free of disease, but are also born free of genetically transmitted physical and mental disabilities.

"Irresponsible parents-to-be will continue to have children of natural conception and risk having children born blind and with other physical and mental impairments, many to suffer abbreviated lives and die unspeakable deaths.

White Light Dark Night

Associated Press

"What is important, is not how many babies are born - but that every child that is born has an equal chance at a good and healthy life!"

"Eventually, men and women of good conscience will step in and natural conception, being cruel and irresponsible, will become illegal. This will, in time, remove intercourse as the acceptable method of procreation. Sexual acts will be sexual acts, a means of intimate communication between people who love or are attracted to each other. When intimate communication is the objective, one will resort to contraception; when bearing children is the objective, one will see one's doctor."

Summation

The Pope summarized his declaration, "Mother Church will take the reigns of the carriage that will lead us into a better world. She will take upon her shoulders to control the world's population and what's more she will work together with the world of scientific research which will eventually bring about a day when all children will be born into the world with an equal chance at a good and healthy life."

At the end of the day, among two or three dozen others, the Pope had issued a few colossal decrees. Some so great in their implications as to uproot centuries of theological assumptions.

He had replaced the doctrine of *Humanae Vitae* with the doctrine of *Planned Parenthood*. He had imposed a vow of poverty on all members of the clergy and ordered the liquidation of the Vatican treasures. He had declared sex in itself no longer sinful. He had declared hatred of one's fellowman to be the greatest of sins. Perhaps, most revolutionary of all, he had allied the Church with science; the two would work together to bring about a day when all children would be born into a world of equal health and opportunity.

He had left one more on the table. He had set aside the *Doctrine of the Holy Trinity*. Yet, he had left for the others to define the true God. Is it the *God of fear and hatred* of Moses or is it the *God of love and compassion* of Mark, Matthew, Luke and John? Above all, he had left no room for them to have their cake and eat it too. He made it clear they could no longer serve a mishmash of two Gods, *"We will make a choice, and we will make that choice now."*

A new beginning

In closing, John Paul repeated the summary statement of his acceptance speech of a few weeks earlier, *"... Let our differences mold into one and together we shall rise to bring the world to a condition of greater justice. We call upon all of you, from the humblest who are the underpinnings of nations, to heads of state. We encourage you to build an efficacious and responsible structure for a new order, this one more just and honest for all."*

His eyes roamed about the field of white smocks which engulfed him as if to give recognition to each and everyone who was privileged to be there. He clasped his hands and bowed his head in a closing prayer.

"Dear Jesus,

"We offer you no great cathedral, no chant, no offering of gold. All we have to give is our promise we will do what you have asked us to do. What you have left behind in the minds of men.

"We know not where this path leads. All we know is that you have told us to take it, and that is all we need know. If at its end there is nothing there, it is enough for us that you have allowed us the opportunity to have walked this way. That you have called upon us to bring about the equality of all your children.

"We will always cherish having been given this great privilege. We care not if it takes us over the highest mountains, or across the widest seas, or against the armaments of all the armies of the earth, or even through fire. We intend to do this thing with all the strength, vigor and courage that is within us as if the very existence of each and every one of your children depends on us, alone."

Blinking in the eyes of a thousand cameras, the Pope picked up a book, *The Diary of Anne Frank.* Speaking very slowly and decisively - he hesitated between each phrase,

" '...I still believe people are good at heart... as I look up into the heavens I think it will all come right... that, in time, people will find the courage and strength to set aside the prejudices they have been brought into... that people will accept one another as the

equals God created them to be... that their differences will go by the wayside... that, in the end, peace and tranquility will prevail.'"

Finally, John Paul read from two dozen or so military citations for bravery in action from as many different countries of the world. Among them,

"For extraordinary heroism while engaged in military battle without regard to his own safety and risk of life no matter how great, to the betterment of lives of others no matter how small; the Republic of Italy is forever indebted to our eternal friend.

Aldo Moro
March 29, 1965"

The Pope waited for complete silence. His voice soft, yet, carefully articulating each word, it resonated from one end of the magnificent chapel to the other, out onto St. Peter's Square, to the ends of the earth.

"This has been my pledge - - It has been my sacred duty - - as Christ has given me the light to see that duty.
"I have carried this hallowed pledge - - this sacred duty - - with me all of my life.
"I have carried it every day - - every hour - - every moment of my life.
"I have carried it in my mind - - and in my heart - - and in my being - - in my very soul.
"Now, it is time to carry out this solemn promise - - to answer this fervent prayer - - To carry it out for each and every one of them.
"That what they dreamed of - - those things they willed to be - - will come to be - - for each of them - - and for me - -and for you - - and for all humanity!"

White Light Dark Night

CNN reporting
December 17, 2043

It is five o'clock in the afternoon here in St. Peter's Square. A light rain has started to fall and umbrellas are popping up here, there and everywhere. All eyes are on the chimney of the Sistine Chapel which, hopefully, any minute will announce a new Pontiff has been named. We expect our first American Pope.

Within the conclave all remains quiet with each cardinal watching the doors of the inner room where the votes are being tallied. Within that room the count has been completed and the name is known.

Almost immediately white smoke begins to rise from the chimney. A loud roar is heard from the square outside. The doors of the room open and the Vatican Secretary of State, Cardinal Pasquale Amedore, approaches one of the cardinals and asks, "Do you accept your canonical election as Supreme Pontiff?"

The chosen cardinal replies, "I accept."

The Secretary asks, "By what name do you wish to be called?"

The chosen cardinal whispers a name to him. Two cardinals approach with a white smock. The new Pope moves toward the balcony.

Outside, the rain is falling more heavily now. White smoke is bellowing from the Sistine Chapel. All eyes are on the balcony.

At last, the doors open. The new leader of the world's largest congregation appears.

One in the crowd, too short to see, asks another, "American?"

The other responds, "No, English."

"English?"

"Yes, she's English. Louise Brown, the first test-tube Pope!"

Addendum

History of the Holy Trinity

Theology is a form of mythology one accepts as truth. Ancient Greeks viewed their Gods as theology. Modern man views ancient Greek Gods as mythology. It bothers him when the writer uses a capital 'G' in referring to a Greek God. It bothers him, because he knows he is right and the Greeks were wrong.

It all depends on which side of the fence one grows up. If one grows up in the western world, one will be a Christian, Muslim or a Jew. If one grows up in the eastern world, one will be a Hindu, a Buddhist or a Tao. If one grew up in ancient Greece, one accepted Zeus as the King of Gods; He created it all.

The fundamental concept of our existence in the world of science is that something had to be infinite; something started it all. The fundamental concept of our existence in the world of religion is that someone with human consciousness started it all. God is infinite, the definition of God in all the world's religions today. If a God is not infinite, then He or She is not God. The only way around this, is to claim a certain God is one and the same God who created it all; a *Holy Duo*, a *Holy Trinity*, or what have you.

Through the years, theology has, from time to time, been based on known facts and at other times on sheer mysticism. In 1352BC, when the Pharaoh Akhenaten, declared there is only one God, the Sun God, *Aten*, he based his decree on known facts; mysticism had nothing to do with it. It was a known fact in his time, as it is today, the source of all energy on earth was the sun; nothing could exist without the sun.

Despite Akhenaten's rationalization, theology turned back to mysticism just a decade later when a 'ghost' claiming to be God appeared to Moses and promised him the Holy Land. The expert says something that makes no sense at all; everyone believes it.

There is, perhaps, no finer example of this than the *Holy Trinity*. That Christ, who contradicted just about everything His alleged Father had to say, is one and the same God. A concept that defies all known biblical and historical facts; yet, one on which the Christian risks his eternity; a concept that makes absolutely no sense at all. A concept that the 'expert' explains to his customers, "It is simply something we do not understand."

"Know one's enemy..."

As a young priest, Luciani told the Vatican cardinals, *"If one is to be successful in bringing down an enemy, one must know who he is, know where he is coming from, know his armament, know his mental embattlement, how he thinks. . ."*

From an early age, Luciani saw his adversary to be the mysticism of the *Holy Trinity*; which enabled preachers on the *right* to carry on the evil work of the *God of Moses* in the disguise of Christ. It was this same misconception that caused preachers on the *left* to be pulled back by the reigns of Moses in their struggle to bring the reality of Christ into a modern world.

It became his *Ecclesiastical Goal in Life* to separate the love that permeates the gospels from the hatred that permeates the books of Moses, to separate Christ from Moses; to put it more pointedly, to separate Christ from His alleged Father. His enemy was clearly the misconception of the *Holy Trinity*.

Chronology of Scripture relative to the *Holy Trinity*

2500-1500BC   Hindu theology encompassing *three persons in one God* is first reduced to writing. The development of silk in the east enabled text to be reduced to writing much earlier vs. the west where papyrus required large lettering and massive volumes; the reason why easterners had earlier literacy than westerners.

1355BC   Moses is born in Syria

1352BC   Pharaoh Akhenaten comes to power and declares there is only one God, Aten, the Sun God; the first monolithic God in the Western Hemisphere.

1335BC   Moses travels south through the Holy Land to Egypt and learns of the monolithic God.

1337-1300BC   Moses returns to Syria and declares there is one God. His strategy is to incite his people through his stories to take the Holy Land. It is that he places himself as a character in stories which take place before his actual birth that he lived 120 years.

253BC   *Septuagint.* Pharaoh Philadelphius orders Hebrew scripture translated to Greek. No surviving record.

250BC   *Vedas.* The Hindu scripture, the oldest surviving scripture in the world. Dali Lama Palace Tibet.

6BC-27AD   Christ's lifetime.

70AD   Earliest date the *Gospel of Mark* could have possibly been written

85AD   Earliest date the *Gospel of Matthew* could have possibly been written

85AD   Earliest date the *Gospel of Luke* could have possibly been written been

95AD   Earliest date the *Gospel of John* could have possibly been written.

100AD   *East* meets *West.* For the first time Christians and Hindus become aware of each other's God. Earliest use of Arabic #6 & #7 in Hindu Numerals.

110-120AD   Most likely time John wrote his gospel

260AD   *Council of Antioch.* On the presumption Peter and Paul died there, the council decrees that succession must be through Rome. It lists the first popes: Peter, Linus and Clement. It is a biblical and historical fact Peter was never known to have been in Rome.

313AD	Constantine founds the Catholic Church; establishes two seats of authority: Rome and Constantinople
328AD	*Nicene Council* establishes the *Ecumenical Council*, the ruling body of the Catholic world for the next six centuries. It consisted of five bishops: Rome, Constantinople, Antioch, Jerusalem and Alexandria.
	The objective of the *Nicene Council* was twofold: 1) define the God of the Catholic world. 2) establish apostolic succession from Peter to Sylvester, the bishop of Rome under Constantine.
350AD	*Sinaiticus*. Oldest surviving biblical record; most of the New and about half the Old Testament. It does not contain the five books of Moses. It contains the book of *Isaiah* and books of prophecy. In Greek, it is likely a copy of the *Septuagint*. British Library.
400AD	*Vulgate*. St Jerome writes the first Latin Bible. No record has survived. The availability of parchment made it possible for the Bible to be reduced to print.
400-425AD	*Codex Vaticanus*. Most of the New Testament and about half the Old Testament. It does not contain the books of Moses. It contains the book of *Isaiah* and surrounding books. Written in Greek, it is likely a copy of the *Septuagint*. Vatican Apostolic Library.
970AD	*Aleppo Codex*. The oldest partial surviving copy of the Hebrew Bible. It contains fifty-seven percent of the leaves of the Old Testament; does not include the books of Moses. Jerusalem Museum.
1009AD	*Leningrad Codex*. The oldest substantial copy of the Hebrew Bible. St. Petersburg, Russia.
1053AD	Leo IX, bishop of Rome, declares his infallibility and becomes the first Pope as we know him today.

Eastern Church breaks from the Roman Church.

1453AD *Guttenberg Bible.* Oldest known surviving Latin text. Twenty copies, two in the Vatican and the rest scattered around the world in national libraries.

1621AD *King James Bible.* The King James Bible and other modern versions have materially changed original wording of the older surviving texts.

A new God
"this one more just and honest"

 Christians have many times changed their definition of God. One can see this clearly if one follows the progression of the Old Testament into the gospels themselves.
 Before Christ's time, one had only the *God of Moses*. There is no mention in the Old Testament of a *Holy Trinity*. Likewise, there is no mention in the New Testament of a *Holy Trinity*. Yet, the New Testament did leave confusion behind it as to the possible divinity of Christ; that he was the same God as God the Creator.

The role of the gospels in the *Holy Trinity*.

 The twelve disciples, who walked at Christ's side, were Peter, John, Andrew, Phillip, Bartholomew, Matthew, Thomas, Simon, Thaddaeus, Judas and the two James. Both the Church and historians agree that none of these men wrote any of the gospels. That is, the disciples Matthew and John were not the evangelists Matthew and John who wrote their gospels.
 The *Gospel of Mark* was the first gospel written and it could not have been written earlier than 70AD. All history books are consistent as to the earliest dates each of the gospels could have possibly been written. History knows this because each of the gospels speaks of historical events which did not occur until these times; unless, of course, the evangelists could foretell the future.
 If a Gospel speaks of the destruction of Herod's Temple, an event that occurred in 70AD, it could not have been written before

70AD. Herod died in 4BC, allowing two years for the killing of the firstborns, Christ was born in 6BC or earlier.

The *Gospel of Mark*, written more than forty years after Christ's death, is the only gospel that could have possibly been written by a man who could have witnessed Christ's ministry. For this reason, it is the most credible of the gospels.

Son of Man

Mark refers to Christ repeatedly as the '*Son of Man*'. Mark makes no mention of the virgin birth and not once refers to Christ as being God. Until the other gospels were written decades later, the Christian world believed Christ to be a prophet of the God of the Old Testament. The very first Christians had not yet learned of the virgin birth; they did not believe in the divinity of Christ.

As already pointed out, in its relationship to Christ, the most important book of the Old Testament is not *Genesis* with its mythical story of *Adam and Eve* as most believe it to be, rather it is *Isaiah* as it contains the only prophesy of the coming of Christ; the only verse in the Old Testament which links it to Christ.

It is the *God of Isaiah*, not the *God of Moses*, who promises to send His Son to restore eternal life to man which had been taken from him by the *God of Moses*.

It is obvious from Mark's testimony, he was unaware of Isaiah's prophecy otherwise he would have surely included the virgin birth in his gospel, as it is the only testimony in the Old Testament that speaks of Christ's divinity.

The role of the oldest surviving scriptures in the *Holy Trinity*

Unlike the gospels, to set a specific date for the writing of any of the books of the Old Testament one must resort to pure guesswork. The reason being, there has survived no substantial copy of any book of the Old Testament which predates the 4th century after Christ. The oldest surviving copy of any of Moses' stories is the more recent 11th century *Leningrad Codex*. That there exists no record creates an expert's paradise in dating the writing of ancient scriptures. One can say anything one wants to say.

The most substantial record of the old scriptures to survive relative to their timing of being reduced to writing is a letter by the Pharaoh Philadelphius in 253AD which is presently held by the library in Alexandria. Philadelphius orders the translation of the Hebrew scripture into Greek, the *Septuagint.*

From all one knows of their culture, the Greeks added the *God of Isaiah* in their translation. It was Greek ideology and not Hebrew ideology that a man could be the *Son of God* and be born of a virgin. As a matter-of-fact, the Greeks believed this to be true of Homer, Plato, Aristotle and Alexander the Great. Each one had been born of a virgin and was recognized as being the *Son of God*. Alexander was living at the time the *Septuagint* was written.

In ancient Greek mythology, the story of the God Zagreus-Dionysus dates back a thousand years before Moses. In the legend, Zeus, King of Gods, speaks, *"Hail thy offspring who come upon thou without lust... Hail thou who has been counted among thieves and murderers... suffered the suffering ...Thou wilt become God from man and rule eternal life."*

"Hail thy offspring who come upon thou without lust" is synonymous with a virgin birth. It follows the Greeks added the only prophesy in the Old Testament of the coming of Christ. *Isaiah 7. "Behold, a virgin shall conceive and shall bear a son, and shall call his name Immanuel." Isaiah 53. "He hath poured out his soul unto death; and he was numbered with the transgressors; and he bare the sin of many, and made intercession of transgressors,"* and later in *Isaiah, "... shall come to rule all the nations of the earth."*

Unlike the *God of Moses* who was a God of hatred and fear, the *God of Isaiah*, like Christ, was a compassionate God who loved all people equally. This had been the way of the Greeks; the reason they fostered the first democracy. It was not the way of the Jews; they were still stoning unwed mothers and prostitutes to death and sacrificing innocent animals to the *God of Moses*.

That the five books of Moses are missing from both the *Vaticanus Codex* and the *Sinaiticus Codex*, the oldest surviving copies of the *Septuagint*, is powerful evidence they were never included in the *Septuagint* to begin with. The Greeks replaced the *God of Moses* with the kinder *God of Isaiah*, who fostered Greek ideology. It was the *God of Isaiah* and not the *God of Moses* who promised to send His Son to save the world.

That the books of Moses are also missing from the oldest Hebrew translation, the *Aleppo Codex*, yet it does contain *Isaiah*, is telling evidence early theologians tried to eliminate Christ's alleged link to Moses and reestablish it solely to the *God of Isaiah* in order to avoid the confusion the *Holy Trinity* confronts us with today; two very different Gods could be one and the same God..

Nevertheless, it is likely Mark was limited to the Hebrew translation and was unaware of or not able to translate the Greek version. Otherwise he would have included the virgin birth for it is the virgin birth that suggests the divinity of Jesus, the *Son of God*.

That Mark, in writing the first gospel, excluded the virgin birth is compelling evidence the prophecy of Isaiah was not in the Hebrew Bible of his time; or he certainly would have included it. One must keep in mind that Mark was in Jerusalem; he would have had access to the most complete Hebrew Bible of his time.

Son of God

A quarter century after the *Gospel of Mark*, comes the *Gospel of Matthew* followed a few years later by *Luke*. The *Gospel of Matthew* is a direct plagiarism of the *Gospel of Mark* which adds the concept of the virgin birth. That Matthew could translate the Greek *Septuagint* and Mark could not, is consistent with the gospels which speak of Matthew as *"one of many tongues."*

It is that Matthew begins his gospel with the birth of Christ, and Mark does not, that modern bibles begin with *Matthew*.

The *Gospel of Luke,* in turn, is a plagiarism of *Matthew* which adds the concept of the *Ascension*. Modern versions of *Luke* add commentary concerning life including the sinfulness of sex. Except for these modern add-ons, Christ never speaks of sex.

Both Matthew and Luke refer to Christ as the *Son of God*, Greek ideology, a step up from Mark who refers to Him as the *Son of Man*, Hebrew ideology. Matthew and Luke copy most of Mark's work word-for-word changing 'Son of Man' to 'Son of God'.

Nevertheless, until the *Gospel of John* was written in the second century, Christians continued to believe in one God - the *God of Moses* with the additional provision, Christ was an emissary of God; the *Son of God*; not necessarily God. The Holy Ghost continued to be viewed simply as a synonym of God.

East meets West

At the turn of the second century east met west. As one knows, Christ at His time was not aware of the Hindus in the east; He was not aware of the existence of another God on the other side of the world. Early in the second century, John the evangelist became aware there could be more than one person in God.

The Hindu 'crucifix' two thousand years before Christ, as it remains today, a single figure with three faces.

Brahma   Vishnu   Shiva

God Brahma, the Creator

God Shiva, the Redeemer

God Vishnu, the Enlightener

Three persons in one God, the Hindu *Trimurti*, the *Holy Trinity* of the *Vedas* dated a thousand years before Moses time; a *Holy Trinity* with identical relative functions the *Nicene Council* would adopt for Christianity centuries after Christ's time. The following relief is taken from Saint Dennis Basilica in Paris

White Light Dark Night

Father, Son and the Holy Ghost
(lamb)   (dove)

God the Father, the Creator

Christ, the Redeemer

Holy Ghost, the Enlightener

In the *Gospel of Mark*, three times Christ denies He is God. Mark 10, *"Jesus said unto him, Why callest me good? There is none good but one, And that one is God."* Yet, just a few decades later, John, who had learned from the Hindus there could be more than one person in God, explicitly declares Christ the same God as the Father, *John 10, "I and my Father are one."*

Christ, in the *Gospel of Mark* - the only gospel which could have been written by a man who could have been a direct witness to Christ's ministry - denies He is God. Yet years later, Christ changes His mind in the *Gospel of John* and declares He is God; a gospel written by a man who could not have possibly been alive at Christ's time.

John, like the other evangelists, describes the Holy Ghost as emanating solely from the Father. None of the gospels describe the Holy Ghost as emanating from Christ. Nowhere in the Bible does it speak of the Holy Ghost as we know Him to be today; a separate person.

There is the biblical and historical fact that John lies in the first verse of his gospel; he claims to be the disciple John *"who bore direct witness to Christ's testament."* History knows he could not possibly have been alive at the time of Christ's ministry.

Nicene Council

Regardless, from the middle of the 2nd century until the 4th century, Christians believed in a *Holy Duo*, Christ and His Father, two divine persons with the Holy Ghost a synonym of the Father.

In 325AD, the *Council of Nicea* met under Constantine. Its objective was to redefine the Christian God. After three years, this body of three hundred bishops came to the conclusion there were two persons in God and that the Holy Ghost emanated individually from both the Father and from the Son. For the next twenty years, this revised *Holy Duo* reigned over the Christian world.

In 345AD a follow-up council met which defined the *Holy Trinity* as we know it today - three equal Deities in one God; three Persons in one God; identical in function to their counterparts of the Hindu God on the other side of the world.

Mythical Succession

In addition to establishing the *Holy Trinity*, the *Nicene Council* established succession from Peter to Sylvester, the sitting bishop of Rome. There had been no organized church in Rome until Constantine, just a few scattered Christians hiding in the catacombs. The Council listed three dozen 'bishops' of Rome, between Sylvester and Peter, only seven of which are known to have lived. The others were mythical creations of the Council.

They spent months fabricating stories of their fictitious lives placing most of them as martyrs in the coliseum and canonized all of them saints. It is this mythical *communion of saints* from Christ to Sylvester, not to saints as we know them today, a Catholic makes his or her pledge of allegiance to his Church - a Catholic's make-believe link to Christ.

The Apostles Creed

I believe in God, the Father Almighty, Creator of heaven and earth; and in Jesus Christ His only begotten Son, Our Lord.

Who was conceived of the Holy Spirit, born of the Virgin Mary, suffered under Pontius Pilate, was crucified, died, and was buried.

He descended into hell.

The third day he rose from the dead.

He ascended into heaven sits at the right hand of God the Father Almighty, whence he shall come to judge the living and the dead.

I believe in the Holy Catholic Church, the communion of saints, the forgiveness of sins, the resurrection of the body, and life everlasting.

Amen

"The fundamental definition of God is 'God is infinite'. '...*Jesus Christ, His only begotten Son...*' Christ was not infinite; He was the *Son of Man* as Mark had said He was.

"Yet, what is important of Christ's life is not His life on earth, not the miracles He is said to have performed, not His death on the cross, not even His ascension into heaven. Like any God, or for that matter, like any man, all that counts is what He left behind,

'Love thy Neighbor as thyself . . . all creatures, great and small.'

"I first said it in 1973,

'Never be afraid to stand up for what is right, whether your adversary be your parent, your teacher, your peer, your politician, your preacher, your constitution, or even your God!'

"No man more practiced it than did Jesus Christ, Himself. That alone, makes Him God.

"Take this with you wherever you go. Yes, one thing more. Each time the fork in the road comes up, often only minutes apart, ask yourself, *'Now, what would Jesus have done in this case?'*

"Do this for me and one day we will enjoy the wine of the Gods at the corner wedge café in Vittorio Veneto together with Jack and Lucien and all the others. We will walk together in the woods with the cat and the fox and the poodle Medoro."

Piccolo, The Mud in the Street

Author

George Lucien Gregoire was born in New England and completed his undergraduate and graduate work in Massachusetts schools. A war veteran, he spent most of his professional life as an officer of corporations in the United States and Europe. For a time, he was an international figure in cooperative education and has served on the boards of universities and secondary schools. He is the founding trustee of a score of charitable organizations, some providing education to mentally and physically impaired children. Gregoire first met John Paul in 1968 when the Pope was a little known bishop of a remote mountain province in Northern Italy.

White Light Dark Night and its sequels *Murder in the Vatican* and *A God for Lions* are previewed and available: www.authorhouse.com www.amazon.com www.barnesandnoble.com and other sites.

Autographed books: See author's listings 'vaticandirect' under 'used & new' on amazon.com or contact Vatican@att.net or 410 625 9741.

Murder in the Vatican contains both books: 1) *The Revolutionary Life of John Paul* and 2) *The Investigation into the Vatican Murders of 1978*. Warning: Do not purchase editions of *Murder in the Vatican* that predate 2006 as they are partial editions. *A God for Lions,* also published as **The Reincarnation of Albino Luciani,** is a collection of short stories based on Luciani's doctoral thesis *Origin of the Human Soul* in which he explores the world's religions and defines the human soul.

Recommended sites:

www.JohnPaulI.org for first chapters of all of the author's books

www.MurderintheVatican.com

web design: www.bgjh.nl

cover design: http://Roberto.setimadimensao.com

www.LiberalsLikeChrist.org

www.defenbaugh.org

All comments appreciated: vatican@att.net 410 625 9741

Lightning Source UK Ltd.
Milton Keynes UK
UKOW042001110113

204787UK00001B/54/A